Analog Circuits and Signal Processing

For further volumes:
http://www.springer.com/series/7381

Pipelined ADC Design and Enhancement Techniques

Consulting Editor: Mohammed Ismail. Ohio State University

Imran Ahmed

Pipelined ADC Design and Enhancement Techniques

Springer

Dr. Imran Ahmed
Kapik Integration
192 Spadina Ave.
Toronto M5T 2C2
Suite 218
Canada
imran@kapik.com

ISBN 978-90-481-8651-8 e-ISBN 978-90-481-8652-5
DOI 10.1007/978-90-481-8652-5
Springer Heidelberg Dordrecht London New York

Library of Congress Control Number: 2010920320

Cover design: WMXDesign GmbH, Heidelberg, Germany

Printed on acid-free paper

Springer is part of Springer Science+Business Media (www.springer.com)

Preface

Pipelined ADCs have seen a tremendous growth in innovation and scope over the past few years. As such understanding both the basic concepts and the leading edge techniques required to realize pipelined ADCs which meet the challenging specifications of today's market and applications is required. While pipelined ADCs are popular circuit blocks, beyond publications in periodicals there are only a few condensed resources which are dedicated to education in the area. This book aims to help bridge the gap with a thorough discussion of pipelined ADCs.

This book is targeted to both the beginner and expert looking to acquire knowledge in pipelined ADCs. In the first section of this book, a tutorial discussion of several key design tradeoffs involved in designing a pipelined ADC is given. The discussion is presented with sufficient detail so as to allow those with only introductory knowledge of pipelined ADCs to quickly understand the limiting factors which motivate research into methods which enhance the performance of pipelined ADCs. In the second half of this book a detailed overview and discussion of four state-of-the-art pipelined ADCs with silicon implementations and measured results is given. The innovations include: a technique to rapidly digitally correct gain + DAC errors in a pipelined ADC, an architecture to enable a single ADC to be designed to achieve low power for a very wide range of sampling rates, a circuit technique to eliminate front-end sample-and-holds in pipelined ADCs, and finally a very low power pipelined ADC architecture based on capacitive charge pumps.

The innovations presented in this book provides several tools which can be of great use to help a pipelined ADC designer deliver a design with good linearity, broad application, and very low power.

Acknowledgements

Research is a unique proposition. One is forced to look into the depths of the unknown and find an answer to a question that does not necessarily have an answer. In some cases your answer fits the question – in some cases your answer fits the question like a square peg in a round hole. Regardless of the madness, the journey of developing abstract ideas into ultimately something which works is truly a unique and completely enriching experience – an experience that I for one am tremendously thankful for and very fortunate to have undergone in developing the material for this book. Acknowledging specific people in the development of an abstract piece of art is somewhat partial, as undoubtedly every person one interacts with during the course of a writing a book in some shape or form impacts the work. There are a few key people however who have helped this work take form. Firstly I must thank Professor David Johns at the University of Toronto. His guidance in developing many of the ideas discussed in this book were invaluable. I am also thankful to Professor Ken Martin, also of the University of Toronto, whose rigor and boldness significantly helped this work take shape. I also thank the support of the team of excellent designers at Broadcom Netherlands, especially Jan Mulder and Klaas Bult, who in addition to providing a wealth of knowledge, have inspired me to be excited about the future in mixed signal circuit design. Of course one cannot accomplish anything in life without the unquestioned pillar of support one's family offers. This book is dedicated to my family.

Contents

Abbreviations

ADC	Analog to digital converter
CBSC	Comparator based switched capacitor
CM	Common mode
CMFB	Common mode feed back
CMPS	Current modulated power scaling
CMR	Common mode rejection
DAC	Digital to analog converter
dBFS	dB relative to full scale
DLL	Delay locked loop
DNL	Differential non-linearity
DNW	Deep N-well
ENOB	Effective number of bits
FOM	Figure of merit
IC	Integrated Circuit
INL	Integral non-linearity
KCL	Kirchhoff's current law
MDAC	Multiplying digital to analog converter
MIM	Metal-insulator-metal
NM	Nominal mode
PGA	Programmable gain amplifier
PRM	Power reduction mode
S/H	Sample-and-hold
SAR	Successive approximation register
SFBO	Switched feedback biased Opamp
SFDR	Spurious free dynamic range
SNR	Signal to noise ratio
SNDR	Signal to noise plus distortion ratio

List of Figures

List of Tables

Chapter 1
Introduction

1.1 Overview

The pipelined topology is a popular option for ADCs which require resolutions on the order of 8–14 bits and sampling rates between a few MS/s to hundreds of MS/s. The popularity of the topology can be attributed to its relatively simple and repetitive unit structure, as well as a significant reduction in the number of comparators required to achieve a fixed resolution when compared to other Nyquist-rate data converters such as Flash, and Folding + Interpolating based converters. Pipelined ADCs are used in a variety of applications such as: mobile systems, CCD imaging, ultrasonic medical imaging, digital receivers, base stations, digital video (e.g. HDTV), xDSL, cable modems, cellular base stations, and fast Ethernet. Since pipelined ADCs are used in a variety of electronic systems, research in improving the performance of pipelined ADCs has attracted much attention over the past decade, where the most popular areas of research have been: linearity enhancement, and power reduction. In recent years, an emerging area of research in ADCs has been the development of reconfigurable ADCs [1].

Linearity enhancement has been an active area of research as with deeper sub-micron technology low intrinsic gain, low supply voltages, and device mismatch have made achieving very linear data converters (i.e. >10-bit linear) challenging using conventional pipelined ADC design techniques. Low power consumption in pipelined ADCs is motivated by the fact that low power consumption enables increased battery life and thus increased user productivity in mobile systems. In wired systems where many ADCs can be integrated on-chip in parallel, large net power consumption can generate high amounts of heat requiring expensive packaging for heat dissipation; hence lower power enables more cost-effective packaging. With the green-shift of modern electronics, a paradigm of 'doing more with less' has become a popular mantra in the latest semiconductor systems. In the interest of saving power as well as recycling as much area in an electronic implementation as possible (and thus reducing implementation costs), reconfigurable

I. Ahmed, *Pipelined ADC Design and Enhancement Techniques*,
Analog Circuits and Signal Processing,
DOI 10.1007/978-90-481-8652-5_1,

data converters, which can operate at a variety of different operating points have become an emerging area of research. A reconfigurable ADC for example which has its power scale with sampling speed, would allow a single ADC to be used to meet the demands a variety of different standards and/or inputs without using multiple ADCs. As multiple devices and communication protocols are integrated onto single devices such as cell phone, designing flexible reconfigurable electronic systems is emerging as an area of great interest to ADC designers.

This book aims to be of interest to those who are new to pipelined ADCs and those who are seasoned experts in the field. The book is divided into two sections: Section I discusses pipelined ADC design, and Section II discusses pipelined ADC enhancement techniques. Although many topics related to pipelined ADCs are discussed in both sections, the primary focus of the book is on design techniques which (1) improve linearity, (2) enable reconfigurable ADCs, and (3) Facilitate low power consumption. Throughout the book examples of prior art and silicon implementations of state of the art solutions in these areas will be described in detail.

Section I of the book is aimed at those new to pipelined ADCs, where a review of the basic knowledge and design trade-offs required to understand how a pipelined ADC works are detailed. By understanding the conventional approaches used to implement a pipelined ADC it is expected that the casual reader can put together a functional ADC with reasonable specifications. By understanding the conventional approaches used to implement pipelined ADCs the reader will gain an appreciation of the limitations of the conventional approaches, and the motivation of state of the art designs described in the second portion of the book.

The second portion of the book is aimed at those already familiar with the basics of pipelined ADCs. In the second portion of the book four state of the art pipelined ADC designs are reviewed and presented in detail. For each implementation a state of the art architectural innovation which substantially improved on the prior art, with verification in silicon will be discussed. The following are brief summaries of the four pipelined ADC architectures discussed in this book:

1. A topology [2, 3] to rapidly digitally correct for both DAC and gain errors in the multi-bit first stage of an 11-bit pipelined ADC. Using a dual-ADC based approach the digital background scheme is validated with a proof-of-concept prototype fabricated in a 1.8 V, 0.18 μm CMOS process, where the calibration scheme improves the peak INL of a 45 MS/s ADC from 6.4 to 1.1 LSB after calibration. The SNDR/SFDR is improved from 46.9/48.9 dB to 60.1/70 dB after calibration. Calibration is achieved in approximately 10^4 clock cycles.
2. A 10-bit pipelined ADC which has its average power is scalable with sampling rate over a large variation of sampling rates [4, 5]. Fabricated in a 1.8 V, 0.18 μm CMOS process, the ADC uses a novel Rapid Power-On Opamp to achieve power scalability between sampling rates as high as 50 MS/s (35 mW), and as low as 1 kS/s (15 μW), while achieving 54–56 dB of SNDR (at Nyquist) for all sampling rates. A current modulation technique is used to avoid the less accurate simulation, poorer matching, and increased bias sensitivity associated with

weakly inverted transistors. The Rapid Power-On Opamp due to its short power on/off time also affords reduced power consumption in high speed pipeline ADCs, where opamps can be completely powered-off when not required. Measured results show a reduction in power by 20–30% when using a Rapid Power-On Opamp in a high speed pipelined ADC.

3. A sub-sampled pipelined ADC architecture which uses the power scalable architecture introduced in [4, 5], but with a further ∼25% reduction in power by using an innovative technique to eliminate the power hungry front-end sample-and-hold [6, 7]. A method to improve the settling behavior of Rapid Power-On Opamps is also developed. Measured results in a 1.8 V 0.18 μm CMOS process verify the removal of the front-end sample-and-hold does not cause gross MSB errors for input frequencies higher than 267 MHz. With $f_s = 50$ MS/s, for $f_{in} = 79$ MHz the SNDR is 51.5 dB, and with $f_s = 4.55$ MS/s for $f_{in} =$ 267 MHz the SNDR is 52.2 dB.
4. A very low power 10-bit pipelined ADC [8] which replaces power hungry opamps with source followers and a differential capacitive charge pump configuration. Foreground calibration is used to compensate for errors that arise from stage-gains that are not exactly 2. Measured results from a prototype operating at 50 MS/s fabricated in a 1.8 V, 0.18 μm CMOS process show the ADC to achieve a peak SNDR/SFDR of 58.2 dB/66 dB at 50 MS/s while only consuming 9.9 mW, for a figure of merit of 0.3 pJ/step.

After reading Section II of this book the reader will be equipped with in-depth knowledge of solutions to many of the key issues a designer faces when designing a state of the art pipelined ADC.

1.2 Chapter Outline

1.2.1 Section I: Pipelined ADC Design

- Chapter 2 provides an overview of ADC converters with an emphasis on Nyquist-rate data converters. The various tradeoffs between different ADC topologies are outlined in the chapter.
- Chapter 3 provides a detailed discussion of the architecture of and circuit design tradeoffs required in designing a basic pipelined ADCs.
- Chapter 4 discusses issues related to reconfigurable design, namely having a power which scales with sampling rate. Design issues such as biasing transistors in weak inversion and the associated implications are reviewed.
- Chapter 5 surveys state of the art pipelined ADC enhancement techniques, where topics most relevant to this book, namely linearity enhancement, reconfigurable ADCs, and low power techniques are detailed.

1.2.2 Section II: Pipelined ADC Enhancement Techniques

- Chapter 6 discusses a novel technique to rapidly measure and correct both DAC and stage-gain errors in the first stage of an 11-bit 45 MS/s ADC. In the chapter the architecture, circuit design, and measured results of a prototype fabricated in 0.18 μm CMOS are presented.
- Chapter 7 discusses a reconfigurable ADC which has a power which scales with operating speed, over a very wide range of sampling rates (1:50,000). The architecture, circuit design, and measured results of a prototype fabricated in 0.18 μm CMOS are presented in the chapter.
- Chapter 8 discusses a new technique to eliminate the front-end sample-and-hold, and thereby enable a large reduction in the power consumption of a pipelined ADC. The architecture, circuit design, and measured results of a prototype fabricated in 0.18 μm CMOS are presented in the chapter.
- Chapter 9 deals with a novel technique to significantly reduce pipelined ADC power consumption by replacing opamps with source followers and a capacitive charge pump technique. In the chapter, the architecture, circuit design, and measured results of a prototype fabricated in 0.18 μm CMOS are presented.
- Chapter 10 summarizes the book.

Part I
Pipelined ADC Design

Chapter 2
ADC Architectures

2.1 Overview

In this chapter an overview of factors which typically determine ADC resolution is outlined. A brief discussion of popular Nyquist-rate ADC topologies is given, where the topologies most relevant to the focus of this book (besides pipelined ADCs) are discussed with the associated tradeoffs of each topology noted. The goal of this section is to provide a general overview and motivation of ADCs in electronic systems, as well as the system constraints which motivate a particular architecture over the other.

2.2 Factors Which Determine ADC Resolution and Linearity

Digital transmission is the most common form of data communication due to its superior signal integrity in the presence of noise. Digital signals can for example be transmitted at baseband as shown in Fig. 2.1 or modulated by a carrier signal, LO, to a higher frequency as shown in Fig. 2.2.

Different applications have different resolution requirements for the ADC in the receiver. For example, in many wireless receiver systems a receiver is required to resolve a received input with a minimum Signal to Noise and Distortion Ratio (SNDR), which is the ratio in dB of signal power to the power of all harmonics and total noise. The Effective Number Of Bits (ENOB) resolved by an ADC is determined according to the following formula [10]:

$$ENOB = \frac{SNDR - 1.76}{6.02} \tag{2.1}$$

In many applications the power of the received signal can vary significantly (e.g. the distance changing between transmitter and receiver in mobile applications), yet

I. Ahmed, *Pipelined ADC Design and Enhancement Techniques*,
Analog Circuits and Signal Processing,
DOI 10.1007/978-90-481-8652-5_2, © Springer Science+Business Media B.V. 2010

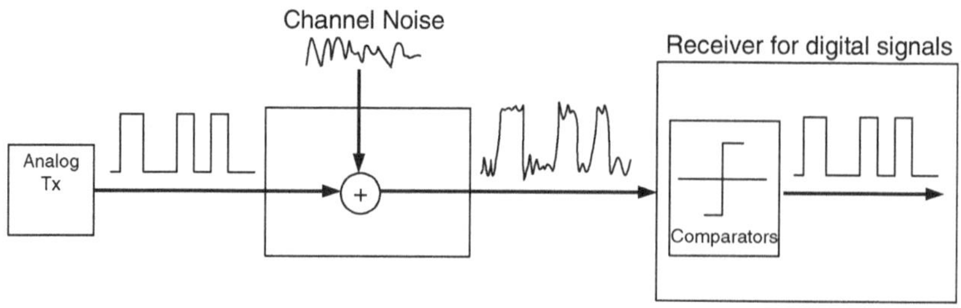

Fig. 2.1 Baseband binary digital transmission

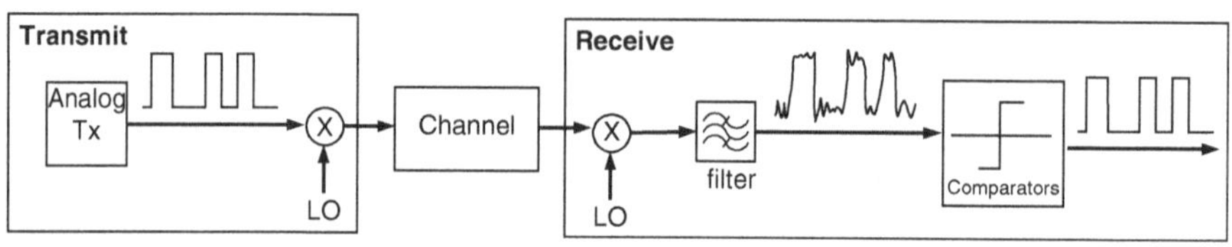

Fig. 2.2 Digital AM transmission

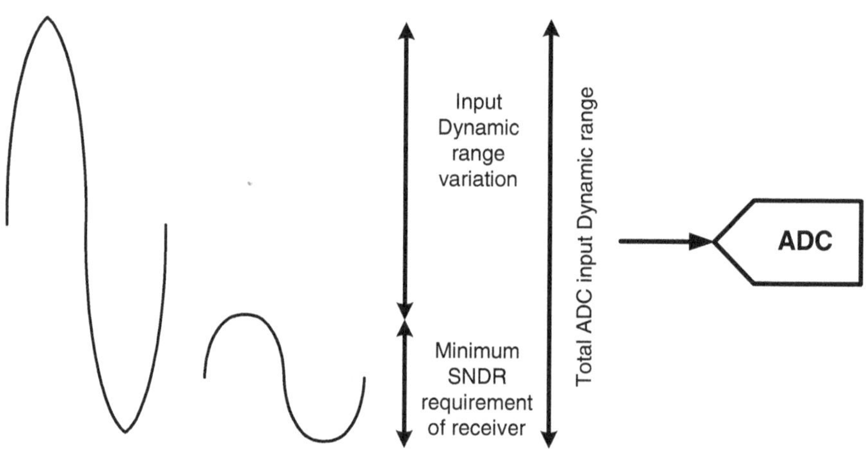

Fig. 2.3 Requirements of ADC input dynamic range

the ADC is still required to linearly digitize the full range of analog inputs with a minimum resolution. In many communication standards, the receive signal is required to maintain a minimum SNDR, thus the total dynamic range of the input (i.e. desired SNDR of the ADC) becomes the product of the minimum SNDR and variation in input signal strength as shown in Fig. 2.3.

In some applications it may be possible to include an Automatic Gain Control (AGC) circuit before the ADC to compensate for the effect of input dynamic range variation, thus reduce the dynamic range hence resolution of the ADC as shown in Fig. 2.4.

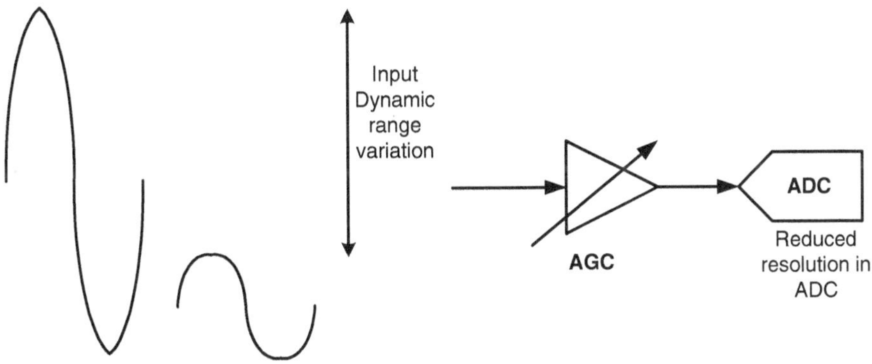

Fig. 2.4 AGC before ADC input to relax ADC input dynamic range requirements

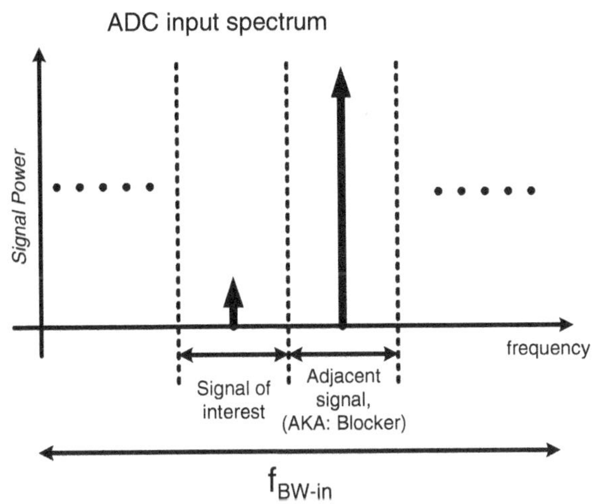

Fig. 2.5 ADC input spectrum illustrating case when desired signal is spectrally next to a much more powerful signal

The required resolution of an ADC can also be determined by the fact that in some applications the signal of interest which is to be digitized is adjacent to another signal in the frequency domain which is orders of magnitude more powerful, as shown in Fig. 2.5.

In such applications it may not be feasible to economically or adequately suppress the adjacent signal (referred to as a blocker) using analog techniques before the ADC input. As a result in many applications the entire signal bandwidth $f_{BW\text{-}in}$ is digitized, and out of band signals efficiently eliminated using digital filters. In such applications to avoid introducing nonlinearity before digitally filtering out the blocker, the ADC must be able to linearly digitize both the large blocker and the signal of interest. Thus the dynamic range requirement of the ADC is set by the sum in dB of the ratio of largest to smallest signal within $f_{BW\text{-}in}$, in addition to the minimum SNDR required to resolve the desired signal of interest.

The required linearity [10] requirement of an ADC can be determined by the fact that in many applications a large number of signals over a large bandwidth are digitized by a single ADC, such as for example OFDM systems, an example spectrum of which is shown in Fig. 2.6.

In an OFDM system the bandwidth of a single signal $f_{sig\text{-}BW}$ is only a small fraction of the overall bandwidth $f_{BW\text{-}in}$ digitized by the ADC. Thus the in-band thermal noise floor for a single signal is very small and given by $P_{NF}f_{BW\text{-}in}/N_{ch}$ (where N_{ch} is the total number of channels), and hence the ADC does not need to be designed with a very low input referred thermal noise floor. If the ADC has a nonlinear transfer characteristic however as shown in Fig. 2.7, each single tone input to the ADC produces an output with the same input tone plus additional harmonics, reducing the linearity in the digital output spectrum. Linearity is commonly assessed in ADCs by taking the ratio in dB of the largest and smallest harmonic tone in-band – referred to as the Spurious Free Dynamic Range (SFDR).

Thus in a system in which the input consists of several frequency adjacent tones such as OFDM, harmonics generated by each unique OFDM signal can fall into the

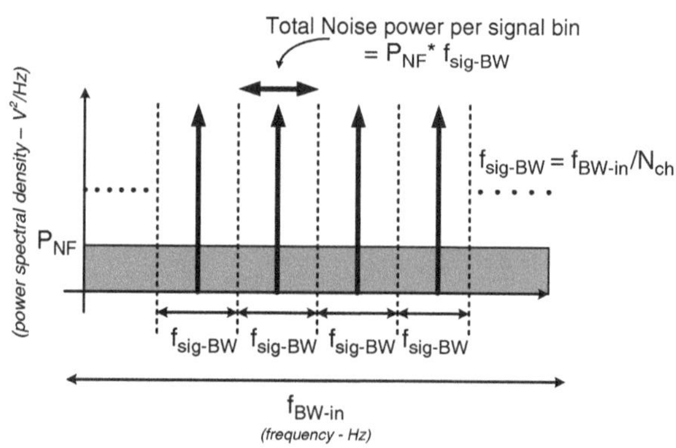

Fig. 2.6 Example OFDM spectrum

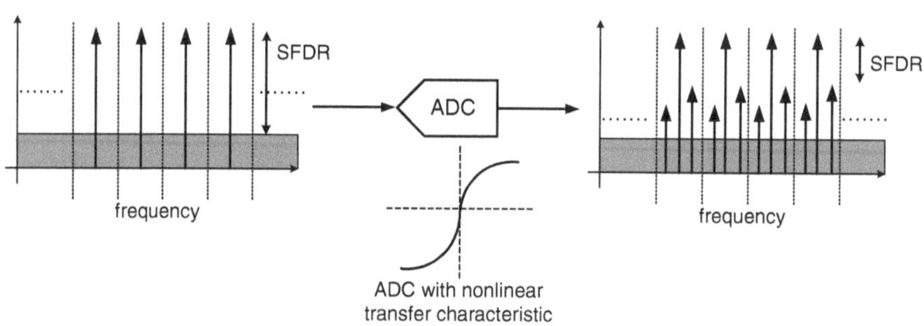

Fig. 2.7 Illustration of harmonic distortion in ADC output spectrum

bandwidth of another OFDM tone, thereby reducing the minimum SNDR within each signal bandwidth $f_{sig\text{-}BW}$. As a result in many systems the linearity of an ADC is the key design parameter, rather than the SNDR. In fact in some publications the ADC ENOB is calculated using the SFDR rather than the SNDR.

From the discussion in this section it is clear that the required resolution and linearity of an ADC can be determined by a number of factors. Furthermore by modifying the receive path by using an AGC and/or a combination of analog filtering techniques, the requirements of an ADC can be relaxed. The optimal configuration for a receiver depends on the specific constraints of a system. It is noted that the general trend in industry is to try and perform as much as possible in the digital domain, thereby pushing the ADC closer to the receiver input, thus demanding ADCs with more resolution and bandwidth – making ADCs essential enablers for future technologies.

2.3 ADC Architectures

As ADCs can consume a large percentage of power in a receiver, it is of vital interest to minimize ADC power consumption. Over the years several architectures have been developed which achieve optimal power consumption for different sampling rates, and resolutions as shown in Fig. 2.8.

The ADC topologies of Flash, SAR, and Pipelined are reviewed in subsequent sections as they are essential to understand within the context of this book. A detailed discussion of topologies not discussed in this book (Delta Sigma, and

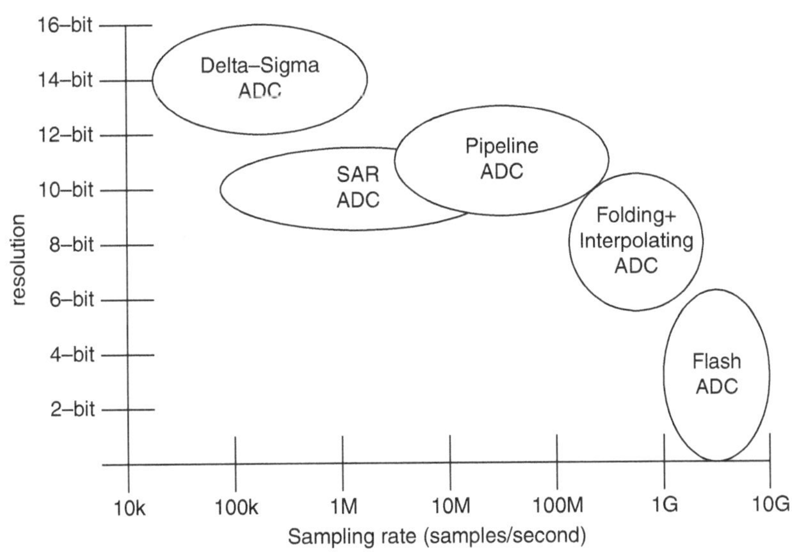

Fig. 2.8 ADC architecture comparison

Table 2.1 Comparison of ADC architectures

Architecture	Latency	Speed	Accuracy	Area
Flash	Low	High	Low	High
SAR	Low	Low-medium	Medium-high	Low
Folding + interpolating	Low	Medium-high	Medium	High
Delta-sigma	High	Low	High	Medium
Pipeline	High	Medium-high	Medium-high	Medium

Folding + Interpolating) can be found in [10]. Table 2.1 summarizes the key tradeoffs of the different ADC topologies shown in Fig. 2.8.

2.4 ADC Figure-of-Merit

A popular Figure-of-Merit (FOM) used to compare different ADCs is

$$FOM = \frac{Power}{(2^{ENOB})(f_s)}\ (\mathrm{pJ/step}) \tag{2.2}$$

where f_s is the sampling rate in Nyquist-rate ADCs. This figure of merit is commonly used to compare published reports as the accuracy term is based on easily measured quantities, and calculates a value that has meaningful units (i.e. energy required per conversion step – thus lower FOM means a better ADC).

A slight variation of equation (2.2) is the following FOM:

$$FOM = \frac{Power}{(2^{2xENOB})(f_s)}\ (\mathrm{pJ/step}) \tag{2.3}$$

In equation (2.3) the ENOB is multiplied by 2 to account for the fact that due to thermal noise limitations, to achieve twice the resolution ×4 the power is required. In general similar FOMs can be achieved with different ADC topologies, however it is noted that ADCs with lower resolutions tend to be able to achieve better FOMs using equation (2.2).

2.5 Flash ADC

Many ADC architectures have been developed over the years, each with different tradeoffs with respect to power, speed, and accuracy. Most ADC architectures however are in some form a variant of the Flash ADC or use a Flash ADC in their implementation.

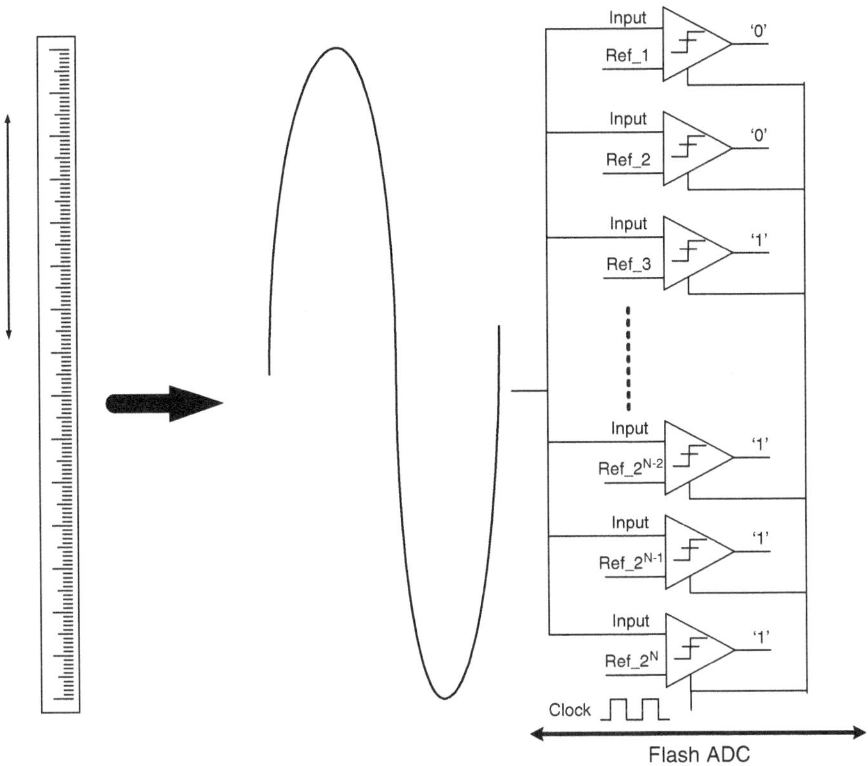

Fig. 2.9 Analogy between a ruler and a flash ADC

Much like a ruler with a fixed resolution maps an infinite precision length to a finite accuracy (e.g. measure a length in millimeters); Flash ADCs *measure* an analog signal into a digital signal by comparing an analog input with fixed reference values as shown in Fig. 2.9. The number of fixed references used determines the accuracy of the digital output. For example, 4-bit accuracy is obtained by comparing against $2^4 - 1 = 15$ reference values, 10-bit accuracy by comparing against $2^{10} - 1 = 1{,}023$ reference values. Determining which reference values the input is in-between forms a length 2^N bit (where N is the accuracy of the ADC) thermometer code representation of the analog input. Mapping the unique thermometer code to its binary equivalent forms a length N binary representation of the analog input [10].

In a Flash ADC, the number of comparators required is thus exponentially related to the desired resolution (in bits). It is noted that to achieve increased resolution in Flash ADCs, large devices are required to suppress process variation effects. Hence in the interest of minimizing area, Flash ADCs are most commonly used in applications where only low resolutions are required.

One of the key advantages of the Flash topology is that it has a potential latency of only one clock cycle – that is the digital output is available within one clock cycle of the input being sampled. In certain systems where an ADC is required in a control loop (e.g. quantizer in a Delta Sigma [10]), it is critical to implement the ADC with as low a latency as possible to maximize closed-loop stability. As low latency is an attractive feature in some systems, many techniques have been developed to enable increased resolution while reducing area consumption, using essentially the Flash topology (e.g.: folding, interpolating, and averaging [10]).

2.6 SAR ADC

The algorithm which forms the basis of the Successive Approximation Register (SAR) ADC has been known since the 1500s, however it was first patented as an algorithm for use in ADCs in 1958 by Bernard M. Gordon [11]. SAR became a popular topology to implement ADCs in the 1970s with the availability of several logic ICs from companies such as AMD. The SAR ADC is relevant to an understanding of pipelined ADCs as from Fig. 2.8 the SAR and pipelined ADC have overlapping areas of use – roughly for low-medium speed applications with 8–10 bits of desired resolution. Thus a clear understanding of each topology and their associated tradeoffs is beneficial in understanding under which circumstances a designer would choose one or the other.

The algorithm used in Successive Approximation is based on a binary search algorithm, and thus is more component efficient than Flash ADCs which use a brute force approach to perform data conversion. Fig. 2.10 illustrates the topology of a standard SAR ADC.

In a SAR ADC the analog input is sampled by a sample-and-hold circuit which operates at the effective Nyquist sampling rate of the ADC, f_s. A sequential binary search is performed on the sampled input by initializing the N-bit register to midscale,

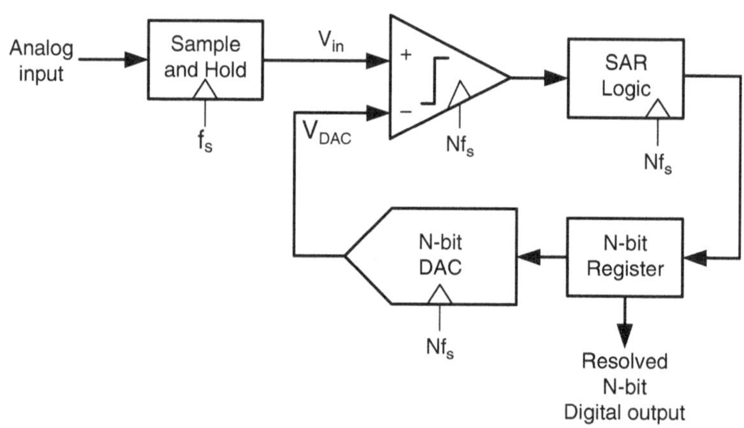

Fig. 2.10 SAR ADC topology

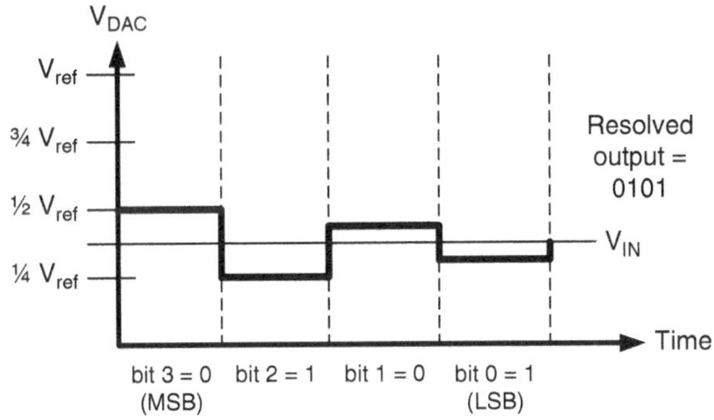

Fig. 2.11 Example SAR conversion – DAC voltage

which forces the decision threshold of the comparator to be $V_{ref}/2$ (where V_{ref} is the full scale input voltage). As a result if the sampled input is greater than $V_{ref}/2$, the MSB of the N-bit register remains at '1', whereas if the input is below $V_{ref}/2$, the MSB of the N-bit register is changed to '0' [12]. By successively repeating the same algorithm and initializing the next bit in the N-bit register to '1', the digital representation of the analog input can be determined to N bits, where N is the resolution of the SAR ADC and the number of times the algorithm is repeated. Figure 2.11 illustrates an example of 4-bit conversion.

The significant advantage of the SAR ADC is that it uses only a few analog components (notably only a single comparator) to implement N-bit data conversion, resulting in a compact area and simple design. Furthermore since the topology produces a new digital output every $1/f_s$, the latency of the ADC with respect to the effective sampling rate is only one clock cycle of the Nyquist-rate clock f_s. As a result the SAR topology can be useful in systems which require ADCs in feedback.

For the SAR ADC to operate with an effective sampling rate of f_s however, the comparator, DAC, and SAR logic are required to operate at Nf_s. For example, if the desired sampling rate and resolution are 100 MHz and 10-bits respectively, the DAC, comparator, and SAR logic are required to operate at 1 GHz. Thus although the SAR ADC allows for a significant reduction in the number of analog components it comes at the cost of restricting the maximum sampling rate to only a fraction of the maximum speed available by a given technology. From a system designer's point of view a SAR ADC may not be feasible in some systems where the available clock is only at the Nyquist sampling rate. As a result SAR ADCs have traditionally been restricted to low to medium speed, and medium to high accuracy applications.

From a design perspective, it is noted that while the SAR topology determines 1-bit of the final digital output every clock cycle, the DAC is required to settle to the full accuracy of the ADC every clock cycle. Also, while any static offset in the comparator appears as an input referred offset to the ADC, the comparator is required to be able to resolve inputs as small as the LSB of the ADC. As a result much effort is required to optimize the DAC and comparator blocks for high speed and high accuracy. It should be noted however that recently published reports [13,14] have begun to emerge which show that low power passive DACs based on charge sharing between capacitors can be used to reduce power consumption and increase the operating speed of SAR ADCs to the point that in 90 nm operating speeds of 50 MS/s can be achieved with resolutions on the order of 8–9 bits while only consuming 0.7 mW [13].

2.7 Sub-sampling

The majority of Nyquist-rate ADCs use a sampling rate which is twice the highest frequency component of the input - this guarantees that the sampled input can be perfectly reconstructed (ignoring the ADC's resolution). In some applications however it is desirable to sample at a fraction of the input frequency – this type of sampling is referred to as sub-sampling. Sub-sampling is commonly used to alias a high frequency input down to a lower frequency. To understand how this happens consider a sinusoidal input *y(t)* of frequency $(f_O + f_{LO})$, which is sampled by an ADC at a sampling rate of f_s. The discrete time values of the input digitized by the ADC are given by:

$$y[m] = \sin\left(2\pi \frac{f_O + f_{LO}}{f_s} m\right) \tag{2.4}$$

If $f_O + f_{LO}$ is greater than f_s, the periodicity of the sine function results in y[m] appearing as if it were sampled at a much lower frequency: modulo $(f_O + f_{LO}, f_s)$. Thus rather than use a mixer to translate an input bandwidth to a lower frequency as shown in Fig. 2.12, sub-sampling can be used to efficiently perform frequency translation as shown in Fig. 2.13. The tradeoff in using a sub-sampled ADC is the maximum input frequency of the ADC is significantly increased.

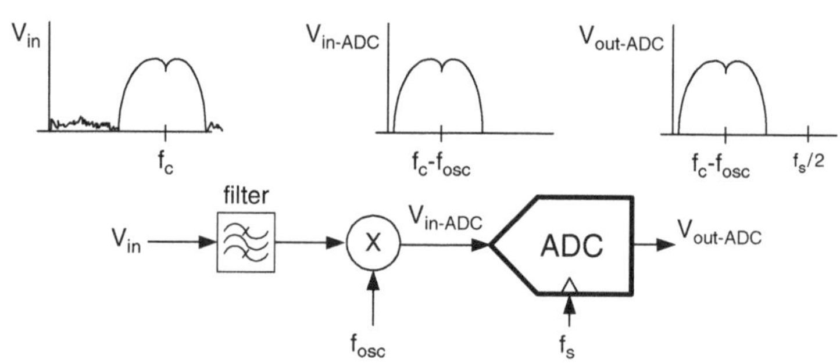

Fig. 2.12 Frequency translation using a front-end mixer

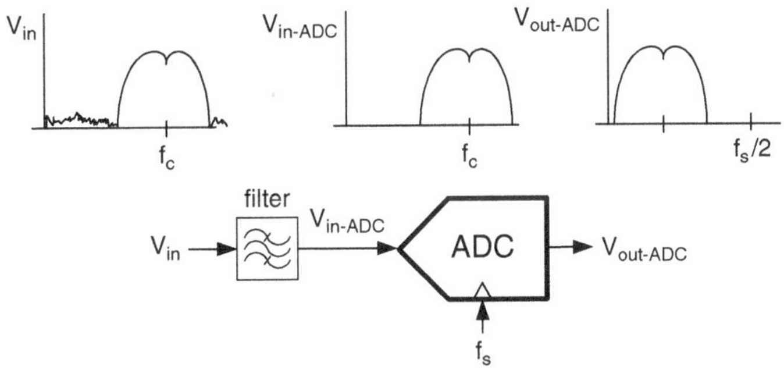

Fig. 2.13 Frequency translation using sub-sampling

2.8 Summary

This chapter an overview of ADC topologies was given. A brief discussion of the system level tradeoffs which determine ADC resolution and speed was given. Nyquist-rate ADC architectures were introduced, with an emphasis of the discussion on the Flash and SAR ADC topologies. The use of ADCs in sub-sampled systems was also described.

Chapter 3
Pipelined ADC Architecture Overview

3.1 Overview

In this section the architecture of a pipelined ADC is introduced, where several key design parameters related to the ADC's resolution and operating speed are discussed in detail. Also discussed in this section is an explanation of how redundancy can be used to significantly relax offset constraints in the sub-ADC of a pipelined stage, as well as the purpose of a front-end sample-and-hold in a pipelined ADC. The goal of this chapter is to provide the reader with an appreciation of the different tradeoffs in designing a pipelined ADC.

3.2 Pipelined ADC Introduction

Pipelined ADCs are capable of resolving medium to high resolutions like the SAR ADC topology, however unlike the SAR topology the pipelined approach is able to achieve very high sampling rates as it does not require a large clock frequency to realize high resolution conversion.

In Fig. 3.1 a two-step ADC or single stage pipelined ADC topology is presented. During the first clock cycle the N/2 Most Significant Bits (MSBs) are resolved (where N is the number of bits in the final ADC output). During the second clock cycle the resolved N/2 MSBs are removed from the input, the remaining signal amplified by 'A' to full scale to maximize the dynamic range, and subsequently the remaining N/2 bits are resolved. Of note, the gain 'A' forms the radix of the digital output, and is often referred to as the 'stage-gain' (a term which will be used frequently throughout this book). The output of a pipeline stage which feeds the input of the subsequent pipeline stage is often referred to as a 'residue'.

Thus the number of comparators required in the two-stage approach is $2^{N/2+1}$, which is lower than the Flash ADC for N > 2. As the pipelined approach implements a queue structure, the maximum speed of the topology is limited by the delay

I. Ahmed, *Pipelined ADC Design and Enhancement Techniques*,
Analog Circuits and Signal Processing,
DOI 10.1007/978-90-481-8652-5_3, © Springer Science+Business Media B.V. 2010

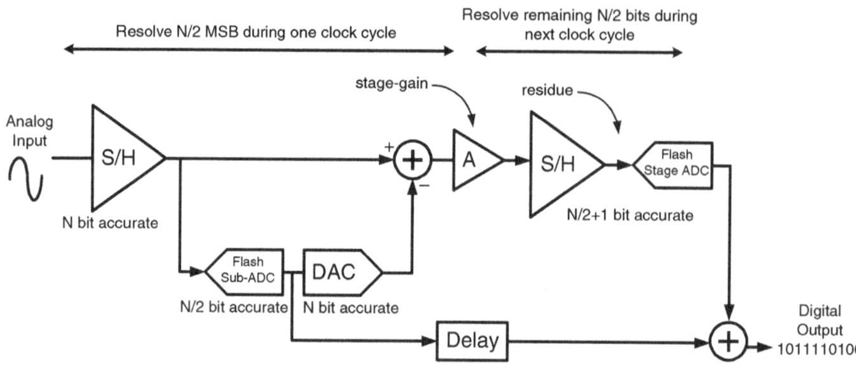

Fig. 3.1 Two step N-bit accurate pipelined ADC

through only a single pipeline stage. However the tradeoff in implementing a queue structure is that the ADC has large conversion latency. That is, rather than the digital outputs being available one clock cycle after the analog input is first sampled (as in the Flash architecture), two clock cycles are required to generate the digital output in the two-step approach.

To ensure the output of the first pipelined stage is sufficiently linear when referred to the input, the *Digital to Analog Converter* (DAC), and subtraction blocks of the first stage must be precise to at least N-bits. Of note however, as the second sample-and-hold is divided by the stage-gain, 'A', when referred to the input, the second sample-and-hold only requires N/2 + 1 bits settling accuracy. In [15] and in Section 3.8, it is shown that by using redundant bits in the sub-ADC, a large offset in the sub-ADC comparators can be tolerated.

The divide and conquer approach used in the two step ADC can be extended further, such that several clock cycles are used, and only a few bits resolved per stage as illustrated in Fig. 3.2; this generalized approach forms the basis of a pipeline ADC.

Although several clock cycles are required for an analog value to be digitized in a pipelined ADC, a new digital output is available every clock cycle. Thus the throughput of the pipeline (i.e. speed) is limited only by the delay through a *single* pipeline stage – unlike a SAR ADC where by virtue of the same stage being reused to resolve the entire digital code the SAR ADC's throughput is limited to a fraction of the comparator clock rate (as described in Section 2.6). Compared to a SAR ADC however, pipelined ADCs require more area as instead of one stage being continuously reused to resolve a digital output, the pipelined approach requires N/n pipeline stages to be implemented on-chip (where 'n' bits are resolved per stage). Pipeline ADCs are useful in configurations where latency is not critical (e.g.) where the ADC is in an open-loop or feed-forward signal path.

The precision requirements of each pipeline stage decrease along the pipeline (i.e.) the first stage must be most precise, subsequent stages need only be as precise as the previous stage less the number of bits resolved previously. Thus analog

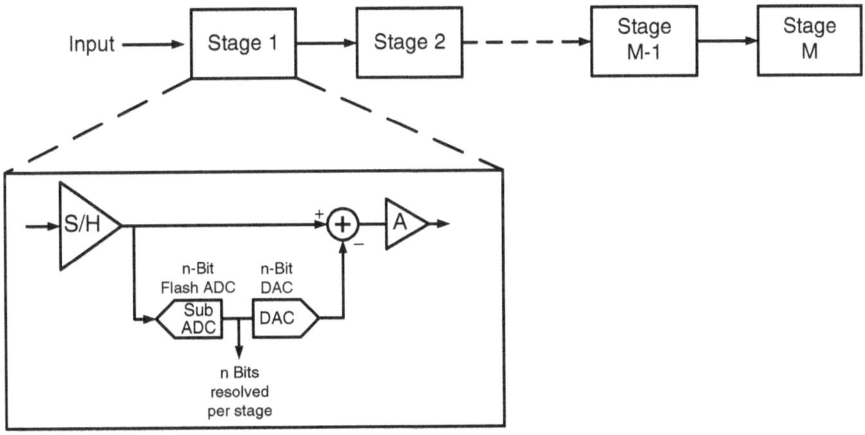

Fig. 3.2 General pipelined ADC architecture

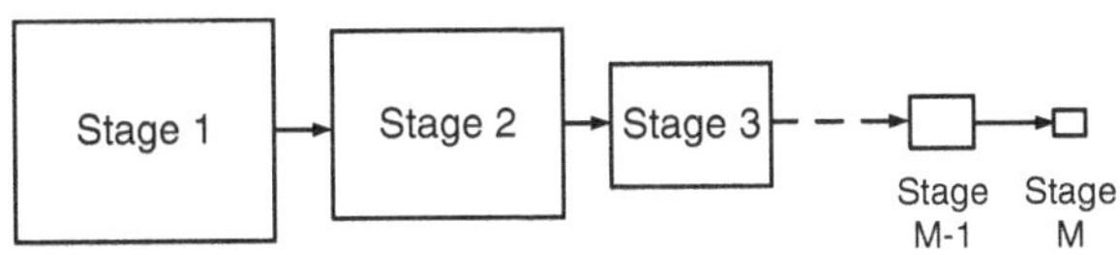

Fig. 3.3 Pipeline stage scaling – stages are sequentially smaller

design complexity can be reduced along the pipeline [16] as shown in Fig. 3.3 (i.e. less opamp gain and bandwidth for later stages – see Section 3.3). As the backend pipeline stages have relaxed precision requirements, they can be designed with smaller area and lower power consumption [16]. Hence it is possible to significantly reduce total power consumption and area by having many stages, where each subsequent stage in the pipeline is sized smaller than the previous stage.

3.3 Multiplying Digital to Analog Converter (MDAC)

In Section 3.2, the topology of a pipelined ADC was described, where it was shown that a pipelined topology was derived from a cascade of pipelined ADC stages, each resolving a fraction of the overall output code. Within a pipelined stage the functions of sample-and-hold, subtraction, DAC, and gain are commonly combined into a single switched capacitor circuit, referred to as a *Multiplying Digital-to-Analog Converter* (MDAC). The MDAC and the sub-ADC make up each pipelined stage in a pipelined ADC as shown in Fig. 3.4.

Figure 3.5 illustrates a generic switched capacitor circuit which implements the functionality of an n-bit MDAC. Opamp based switched capacitor circuits are by far the most commonly used topology to implement the MDAC in both academia

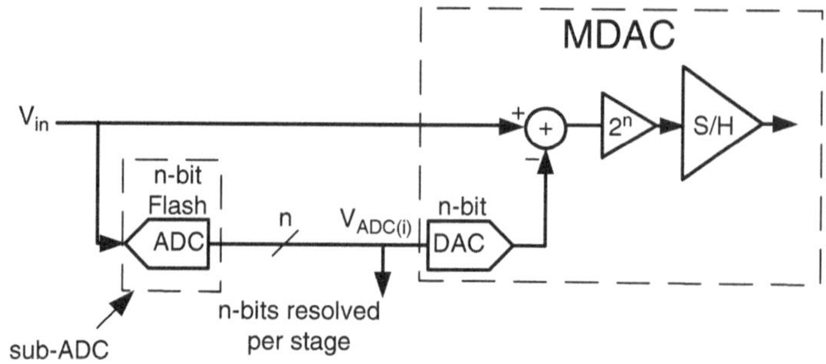

Fig. 3.4 Pipeline stage functionality

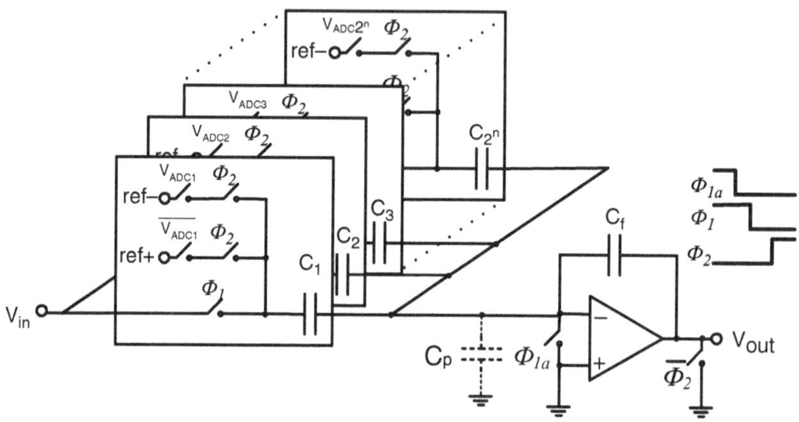

Fig. 3.5 Generic MDAC circuit (shown single-ended but can also be implemented fully-differentially)

and industry. Opamp based approaches enable MDACs which have closed-loop gains that are primarily a function of capacitor mismatch (which can be better than 10-bits), and opamp gain at DC.

From Fig. 3.5, during Φ_{1a} the input is bottom plate sampled on capacitors C_0–$C_2{}^n$. Bottom plate sampling significantly reduces the signal dependent charge injection on the sampling capacitors [10]. During Φ_2, the outputs of the sub-ADC are latched and the MDAC configured to multiply the sampled input by: $\sum_1^{2^n} C_i/C_f$ and perform a DAC operation by connecting one end of each sampling capacitor to either $+V_{ref}$ or $-V_{ref}$ as a function of the sub-ADC output. Assuming the opamp has fully settled, at the end of Φ_2, V_{out} can be expressed as:

$$V_{out} = \frac{\sum_1^{2^n} C_i}{C_f} V_{in} - \left[\frac{\sum_1^{k} C_i}{\sum_1^{2^n} C_i} V_{ref} - \frac{\sum_{k+1}^{2^n} C_i}{\sum_1^{2^n} C_i} V_{ref} \right], \tag{3.1}$$

where k is set by the output of the sub-ADC in the pipeline stage.

3.4 Opamp DC Gain Requirements

To understand the gain requirements of the opamp in an MDAC circuit, consider the hypothetical two stage pipelined ADC topology of Fig. 3.6 which shows a pipelined stage that has its MDAC implemented by switched capacitor circuits, an ideal n-bit sub-ADC, and an ideal backend N-1 bit Flash ADC.

Figure 3.7 illustrates the residue transfer characteristic (i.e. input/output plot for the pipeline stage) of the first stage MDAC when the DC gain of the opamp is infinite, the opamp bandwidth infinite, and no capacitor mismatch. Also shown is the total ADC output which is generated by a summation of the bits generated by the sub-ADC from the first pipeline stage and the backend ideal N-1 bit Flash ADC.

If the opamp of Fig. 3.5 has a DC gain of A, the transfer function from input to output can be found to be:

$$\frac{V_{out}}{V_{in}} = \frac{\sum_1^{2^n} C_i}{C_f}\left(\frac{1}{1+\frac{1}{A\beta}}\right) \tag{3.2}$$

where β is the feedback factor, and is given by:

$$\beta = \frac{C_f}{C_f + C_p + \sum_1^{2^n} C_i}, \tag{3.3}$$

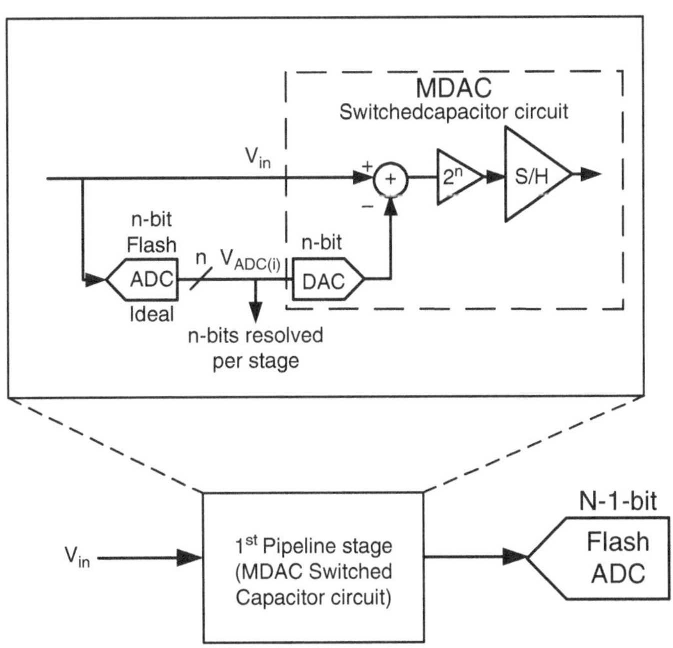

Fig. 3.6 Hypothetical pipeline ADC for illustration purposes

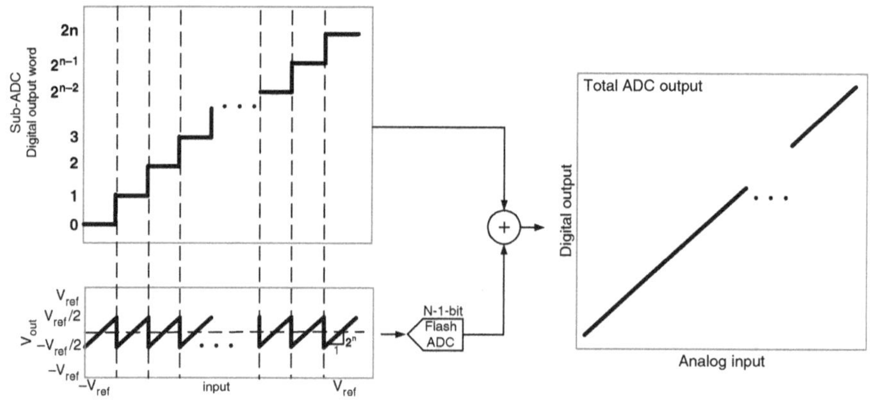

Fig. 3.7 MDAC residue transfer curve, and total ADC output when opamp DC gain is infinite, opamp bandwidth is infinite, and capacitor mismatch ignored

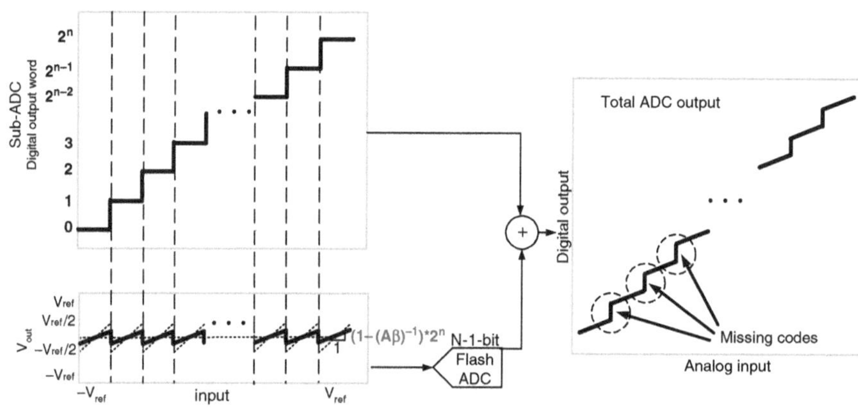

Fig. 3.8 Residue transfer curve of pipeline stage when opamp gain error is included (error free residue curve shown in dashed lines)

and C_p is the parasitic capacitance at the input of the opamp. Thus it can be deduced that the error in the fully settled output, Δ, due to finite DC opamp is:

$$\Delta \approx \frac{1}{A\beta} \quad (3.4)$$

Clearly for a finite A, the gain error is non-zero. Figure 3.8 illustrates the impact on the residue of the first pipeline stage when the effect of finite DC opamp gain is included while ignoring the effect of capacitor mismatch and assuming the opamp output has fully settled, as well as the overall output of the ADC.

From Fig. 3.8 it is clear that finite DC gain in the opamp alters the jumps in the residue transfer curve whenever the sub-ADC output changes, from the ideal value of V_{ref}, to a scalar of V_{ref}. As a result missing codes are generated in the overall pipelined ADC output whenever the sub-ADC output changes. Missing codes in the ADC output result in harmonic distortion which limits ADC accuracy and linearity [10]. To achieve a sufficiently linear pipelined ADC missing codes need to be eliminated. The minimum gain required to achieve no missing codes can be derived by referring the gain error Δ to the input and forcing the total gain error to be below the quantization noise floor of the ADC. For example, for the case of Fig. 3.6, the gain error referred to the input is:

$$\Delta_{input} \approx \frac{1}{A\beta 2^n} \tag{3.5}$$

For an N-bit pipelined ADC, the total sum of errors is required to be less than $1/2^N$. Ignoring all other non-idealities, thus:

$$\frac{1}{2^N} > \frac{1}{A\beta 2^n} \tag{3.6}$$

hence:

$$A > \frac{2^{N-n}}{\beta} \tag{3.7}$$

For example: if N = 11-bits, n = 3-bits, and $\beta = 1/8$, the minimum DC gain in the opamp is ~66 dB to sufficiently suppress the effect of the finite gain of the first stage only. Of note, to a first order $\beta \approx 2^{-n}$, thus equation (3.7) could be also expressed as $A > 2^N$. Clearly a large DC gain is required when opamp based switched capacitor circuits are used to implement the MDAC in medium to high resolution pipelined ADCs. In practice the backend ADC is not ideal and introduces additional non-ideality, forcing the gain of the first stage to be even larger. The net effect of all non-idealities can be found by referring each non-ideality to the input and forcing the sum of all errors to be below the quantization noise floor, i.e. less than 2^{-N}

With modern sub-micron technology nodes offering less and less intrinsic gain in the active region for CMOS transistors (due to short channel effects), attaining opamps with large DC gain can be difficult. Large DC gain is commonly achieved in CMOS using techniques such as gain boosting [21], multi-stage opamps [22], or using long channel lengths for key transistors [10]. However such gain enhancement techniques increase the power of the opamp, and/or increase the design complexity of the opamp. Rather than using analog techniques to suppress the effect of finite DC gain, digital techniques can be used where the error due to finite gain can be compensated in the digital output of the ADC. Such techniques are referred to as calibration and are covered in more detail in Section 5.2.

3.5 Opamp Bandwidth Requirements

In Section 3.4 the DC gain requirements of an opamp based MDAC were derived under the assumption that the opamp had sufficiently settled such that the gain error was limited only by the finite gain error of the opamp. In order to have an opamp which is sufficiently settled within a given timeframe the opamp must have enough bandwidth. As will be seen in this section, an opamp which has a large bandwidth requires large power consumption – thus to minimize power it is critical to optimize opamp bandwidth.

Assuming a first order response of $A(s) = A/\left(1 + \frac{s}{\omega_{3dB}}\right)$ for the opamp of Fig. 3.5 (where ω_{3dB} is the 3 dB bandwidth of the open-loop), near the unity gain frequency ω_{ta}, $A(s) \approx \omega_{ta}/s$ (recalling that $\omega_{ta} = A\omega_{3dB}$ for a first order system [10]). Thus the closed-loop transfer function, H(s), of the MDAC topology of Fig. 3.5 during Φ_2 is approximately given by:

$$H(s) \approx \frac{\sum_1^{2^n} C_i}{C_f}\left(\frac{1}{1 + \frac{s}{\beta\omega_{ta}}}\right) \tag{3.8}$$

Hence the bandwidth of the closed-loop, $\omega_{3dB\text{-}CL}$, is only **a fraction** of the unity gain frequency of the open-loop, i.e.:

$$\omega_{3dB-CL} = \frac{\omega_{ta}}{1/\beta} \tag{3.9}$$

The step response in the time domain of equation (2–10) is given by:

$$h(t) = \frac{\sum_1^{2^n} C_i}{C_f}\left(1 - e^{-t/\tau}\right) \tag{3.10}$$

where $\tau = (\beta\omega_{ta})^{-1}$. The relative error in the output of the first pipeline stage due to finite bandwidth (Δ_{BW}) after T/2 s (where T is the period of clock Φ_2 from Fig. 3.5) is thus given by:

$$\Delta_{BW} = e^{-T/2\tau} \tag{3.11}$$

Referring the settling error of the first pipeline stage to the input of the ADC, and noting that the total error must be less than the quantization noise (i.e. $<2^{-N}$), the required unity gain frequency of the opamp f_{ta} to achieve N-bit settling is thus given by:

$$f_{ta} > \frac{(N - n)\log 2}{\beta\pi} f_s \tag{3.12}$$

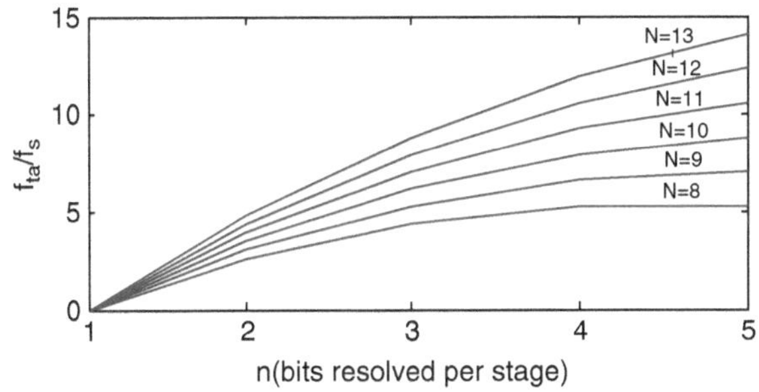

Fig. 3.9 Variation of required unity gain frequency relative to sampling rate with number of bits resolved in the first pipeline stage

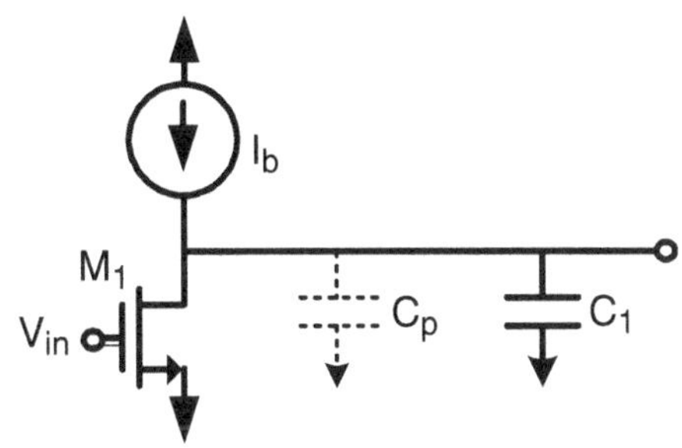

Fig. 3.10 Modeling an opamp by a single transistor

where $f_s = 1/T$ is the sampling rate of the pipelined ADC. From (3) if parasitic capacitance C_p at the input of the opamp is ignored, every time n increases by 1-bit, β ideally decreases by a factor of 2. Using this relationship Fig. 3.9 plots f_{ta}/f_s versus n for different N, where it is seen that for high resolution ADCs as the number of bits per stage n increases the minimum required open-loop unity gain frequency of the opamp to ensure sufficient settling also increases.

If parasitic capacitances are included, β is made smaller, thus even more bandwidth from the opamp is required to achieve the desired settling.

If an opamp is approximately modeled as a single transistor in the active region as shown in Fig. 3.10, it can be readily derived that the unity gain frequency is given by:

$$\omega_{ta} = \frac{g_m}{C_{load}} \tag{3.13}$$

where g_m is the transconductance of M1 and $C_{load} = C_1 + C_p$. From [10], the total parasitic capacitance is given by the sum of drain to gate (C_{dg}) and drain to bulk (C_{db}) capacitances so that:

$$C_p = C_{dg} + C_{db} = WL_{ov}C_{ox} + A_dC_{jd} + P_dC_{j-sw} \tag{3.14}$$

where W and L are the width and length of M1, A_d the area of the drain, P_d the perimeter of the drain, and L_{ov}, C_{ox}, C_{jd}, and $C_{j\text{-}sw}$ process dependent parameters. Since:

$$g_m = \sqrt{2\mu C_{ox}\frac{W}{L}I_D} \tag{3.15}$$

it can thus be seen that:

$$\omega_{ta} = \frac{\sqrt{2\mu C_{ox}\frac{W}{L}I_D}}{WL_{ov}C_{ox} + A_dC_{jd} + P_dC_{j-sw} + C_1}. \tag{3.16}$$

Thus from equation (3.16), in order to double the unity gain frequency, if the width of the device is increased by a factor of 2 (i.e. two devices in parallel), the drain current is required to more than double. This is because as the area of the transistor is increased to increase g_m, the loading of the parasitic capacitor also increases (while C_l remains fixed), and thus reduces the efficiency of the transistor. For large unity gain frequencies – which could be required from an opamp if fast settling is required with a large closed-loop gain (i.e. small β), the parasitic capacitance C_p can be on the order of the load capacitance C_l. Thus in such situations much power is wasted by the transistor in charging both the load capacitance C_l and the transistor's own parasitics. Clearly to avoid such a situation it is beneficial to achieve the maximum speed possible with the lowest unity gain frequency possible – which occurs when the gain, 1/β, is smallest, i.e. when the fewest number of bits are resolved per stage.

In [10] it is also shown that $g_m = 2I/V_{eff}$, where I is the bias current of the transistor, and V_{eff} the overdrive voltage ($V_{eff} = V_{gs} - V_t$). Hence it can also be shown that: $\omega_{ta} = 2\beta I/V_{eff}$. Thus if the gain of the MDAC is increased (i.e. β decreased), to maintain the same closed-loop settling time, the current 'I' needs to be *increased* by the same amount (assuming V_{eff} is made constant). Hence there exists a clear tradeoff between closed-loop gain and power in MDACs (assuming a constant C_l).

3.6 Thermal Noise Requirements

In Section 3.5 it was shown that MDAC settling accuracy was a direct function of the opamp's unity gain frequency. Since the unity gain frequency of an opamp is given approximately by g_m/C_L, the question thus arises what determines the size of the load capacitance? One of the main issues which determines capacitor size is thermal noise. This section introduces the concept of sampled noise and discusses how capacitor size affects sampled noise.

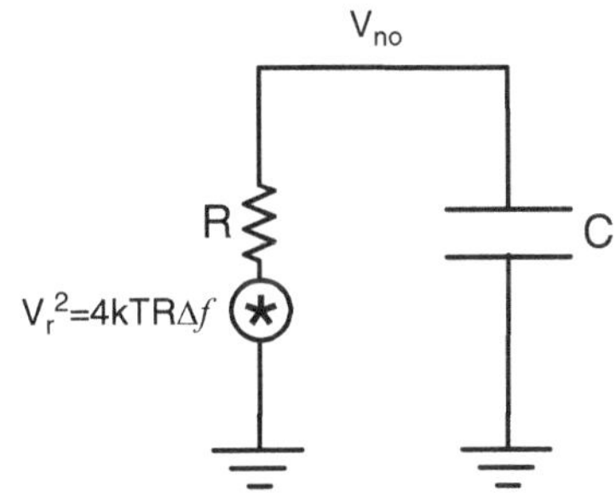

Fig. 3.11 RC noise model

Although capacitors are ideally noiseless elements, in a sampled system, capacitors which hold discrete time values capture noise generated by noisy elements such as resistors (from sampling switches) and opamps. Consider the following noise analysis of a capacitor sampling resistor noise as shown in Fig. 3.11:
from [10] it is shown the equivalent noise bandwidth is $\frac{\pi}{2}f_0$,

$$V_{no-RMS}^2 = \frac{\pi}{2} f_0 V_R^2(f) \tag{3.17}$$

$$f_0 = \frac{1}{2\pi RC} \rightarrow V_{no-RMS}^2 = \frac{kT}{C} \tag{3.18}$$

From the above example it is clear increasing the size of the sampling capacitor reduces the power of thermal noise. As thermal noise represents a dynamic noise source that reduces ADC SNR, sufficiently large sampling capacitors in Fig. 3.5 must be used to suppress the thermal noise of the MDAC to a level below an LSB. Thus thermal noise imposes a tradeoff between power and accuracy – the larger the capacitance (thus power consumption), the lower the thermal noise hence higher the accuracy. Large capacitive loads result in large power hungry opamps. In general the opamp is the dominant consumer of power in pipelined ADCs, thus most power techniques to save power in pipelined ADCs usually involve optimizing the power consumed by the opamps.

In addition to thermal noise from the sampling switches, the MDAC opamps also add thermal noise their outputs. An analysis of the power spectral density of various opamp topologies is given in [23].

3.7 MDAC Design: Capacitor Matching/Linearity

Recalling that the residue of a generic pipeline stage (from Fig. 3.5) is given by:

$$V_{out} = \frac{\sum_1^{2^n} C_i}{C_f} V_{in} - \left[\frac{\sum_1^k C_i}{\sum_1^{2^n} C_i} V_{ref} - \frac{\sum_{k+1}^{2^n} C_i}{\sum_1^{2^n} C_i} V_{ref} \right], \tag{3.19}$$

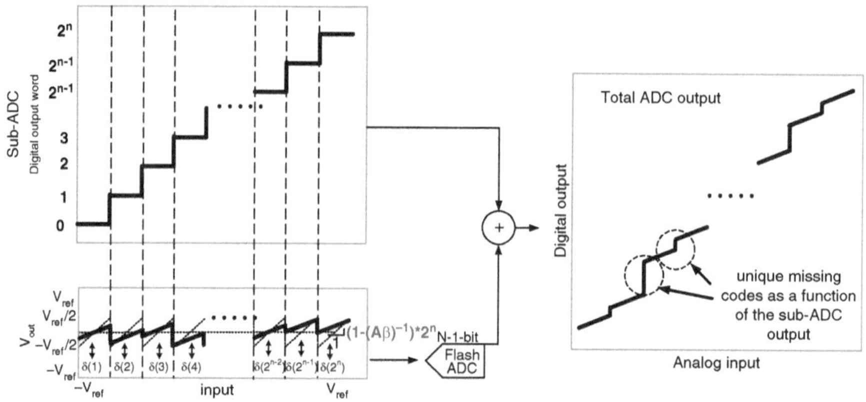

Fig. 3.12 Illustration of DAC and gain errors in pipelined ADC output – ideal residue transfer curve shown by dashed lines

if there is mismatch between $\sum_{1}^{2^n} C_i$ and C_f there will be an error in the desired gain of the pipeline stage, producing similar errors to those illustrated in Fig. 3.8 (i.e. stage-gain errors). Capacitor mismatch *between* sampling capacitors, C_i, also affects the linearity of the DAC in Fig. 3.5. For example, if n = 4 and C_1 = 100 fF, C_2 = 95 fF, C_3 = 98 fF, and C_4 = 103 fF, for k = 1, 2, 3, 4, the amount of V_{ref} subtracted is: −0.4949 V_{ref}, −0.015 V_{ref}, 4797 V_{ref}, and 1 V_{ref}, respectively. For the DAC operation to be error free, every time the sub-ADC output increases by one, the DAC voltage subtracted should increase by 0.5 V_{ref}, however as illustrated with the example, when capacitor mismatch is included this is not the case. The result of these DAC errors is a **unique** jump in the residue transfer function every time the output of the sub-ADC changes as illustrated in Fig. 3.12, resulting in missing codes and substantial harmonic distortion.

The relative mismatch between two capacitors is determined by the area of a capacitor where:

$$capacitor\_mismatch \propto \frac{1}{\sqrt{capacitor\_area}} \tag{3.20}$$

Thus large capacitors are required to achieve a high degree of matching so as to achieve a linear ADC input/output relation. Hence capacitor size is a function of thermal noise *and* capacitor mismatch. As discussed in Section 2.2 in some applications thermal noise requirements are low, however the linearity requirements are high, thus forcing a pipelined ADC designer in such instances to use large capacitors to suppress DAC errors, even though low a low thermal noise floor is not necessary.

Improved capacitor matching in layout can be achieved by using arrays of common unit size capacitors in a highly symmetric configuration, and/or dummy capacitors around the periphery of the capacitor [23]. Capacitor mismatch can also be improved by using Metal Insulator Metal (MiM) capacitors. However, as MiM capacitors require an extra layer during fabrication, they are not always available in processes used to make commercial products, as commercial products tend to be optimized in cost for digital circuits. Due to process variation however, layout techniques and additional design layers can only improve capacitor mismatch so far. Thus an arbitrarily high level of matching cannot be achieved using good layout techniques alone, hence limiting pipelined ADC resolution to the medium-high range as noted in Fig. 2.8. One of the strategies used to overcome limitations due to process variation is to use digital calibration techniques. Calibration techniques can measure and compensate the effect of non-idealities such as finite DC opamp gain, and capacitor mismatch. More details on calibration are provided in Section 5.2.

3.8 Error Correction in Pipelined ADCs: Relaxed Sub-ADC Requirements

The digitization of an analog signal in a pipeline ADC is very similar to the calculation of a quotient in long division, i.e.:

$$\begin{array}{r} Quotient \\ Divisor \,\overline{)\,Dividend}, remainder \end{array}$$

The divisor is similar to the analog input signal (relative to full scale), the dividend the full-scale voltage (i.e. the decimal representation of the largest 10-bit number – 1,023), the quotient is the resolved digital output word, and the remainder the quantization error. By exploiting the long division structure of a pipeline ADC, the accuracy requirements of the sub-ADC can be relaxed. Consider the long division of two numbers: x (divisor), and $y_n y_{n-1} y_{n-2} \ldots y_1 y_0$ (dividend), in an arbitrary but common base β. Both x and y are of arbitrary length, where each digit of y is explicitly shown by the subscripts (most significant digit of y is y_n, least significant digit is y_1). Thus a correct long division of y by x is as follows:

$$\begin{array}{l} \quad\;\; \alpha_n\alpha_{n-1}\ldots\ldots\ldots\alpha_1 \\ x\,\overline{)\,y_n y_{n-1}\ldots\ldots\ldots y_1} \\ \underline{-x\alpha_n} \\ (y_n - x\alpha_n)\beta + y_{n-1} \\ \underline{-x\alpha_{n-1}} \\ \quad r_1 \end{array}$$

r_1 is the remainder after two lines of division.

If however the divisor, x, is incorrectly divided into the dividend, y, an incorrect remainder results, yielding every subsequent digit in the quotient incorrect. This situation is analogous to a pipeline ADC where in a pipeline stage a comparator in the stage Flash ADC, due to an offset, incorrectly sets the stage DAC, leading to an incorrect value being subtracted from the stage input. An important observation is in long division the error is passed to the subsequent line of long division. Thus if a division error could be identified, the error could be eliminated in the subsequent line of long division by adjusting the quotient.

$$\begin{array}{l} \quad\ \alpha'_n\alpha'_{n-1}\cdots\cdots\alpha_1 \\ x\,\overline{)\,y_n y_{n-1}\cdots\cdots y_1} \\ \underline{-x\alpha'_n} \\ (y_n - x\alpha'_n)\beta + y_{n-1} \\ \underline{-x\alpha'_{n-1}} \\ \quad r_2 \end{array}$$

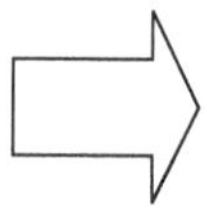

Thus if an incorrect division is made, such that α'_n is an incorrect digit in the quotient, the error can be eliminated by selecting α'_{n-1} such that $r_2 = r_1$

Since the correct and corrected long division approaches yield the same remainder, the quotients in each approach are equal; despite the fact the latter approach included a division error.

The following example numerically illustrates the error correction concept [17]:

Correct division example		Error in division, with correction example	
7.142857		7.143(−2)57	
7) 50.000000		7) 50.000000	
Subtracted reference	−49	Subtracted reference	−49
Residue	1	Residue	1
Amplified residue	10	Amplified residue	10
Subtracted reference	−7	Subtracted reference	−7
Residue	3	Residue	3
Amplified residue	30	Amplified residue	30
Subtracted reference	−28	Subtracted reference	−28
Residue	2	Residue	2
Amplified residue	20	Amplified residue	20
Subtracted reference	−14	Subtracted reference	−21 ← error
Residue	6	residue	−1
Amplified residue	60	Amplified residue	−10
Subtracted reference	−56	Subtracted reference	+14
Residue	4	Residue	4
Amplified residue	40	Amplified residue	40
Subtracted reference	−35	Subtracted reference	−35
Residue	5	Residue	5
Amplified residue	50	Amplified residue	50
Subtracted reference	−49	Subtracted reference	−49
Residue	1	Residue	1
		* Note how error is allowed to pass on to subsequent line of division, and how error is corrected in subsequent line of division	

Correct division quotient:

$$7 \times 10^0 + 1 \times 10^{-1} + 4 \times 10^{-2} + 2 \times 10^{-3} + 8 \times 10^{-4} + 5 \times 10^{-5} + 7 \times 10^{-6}$$
$$= 7.142857$$

Incorrect division with corrected quotient:

$$7 \times 10^0 + 1 \times 10^{-1} + 4 \times 10^{-2} + 3 \times 10^{-3} + (-2) \times 10^{-4} + 5 \times 10^{-5} + 7 \times 10^{-6}$$
$$= 7.142857$$

It is clear a finite error in long division can be tolerated so long as the error passes to the subsequent line of long division, and the occurrence of an error can be detected. Thus error correction can be achieved in a pipelined ADC when errors caused by comparator offsets in the sub-ADC are passed to the subsequent pipeline stage, and a logic implemented to recognize the occurrence of an error.

A simple pipeline topology is one that resolves two bits per stage as shown in Fig. 3.13, the transfer function of which is shown in Fig. 3.14.

The stage gain is ×4 to maximize the dynamic range of the subsequent stage, and to allow for reuse of the reference voltages. An error in the stage ADC threshold (due to an offset) alters the transfer function as shown in Fig. 3.15.

Thus threshold errors lead to stage outputs that exceed the full-scale input to the subsequent stage. As stage inputs that exceed full scale are attenuated or clipped, offset induced errors do not pass to the subsequent stage unaltered, and thus cannot be completely eliminated. If however the stage gain is reduced to ×2 as shown in Fig. 3.16, the error is fully passed on to the subsequent stage, so long as the offset error does not exceed Vref/4, as shown in Fig. 3.17.

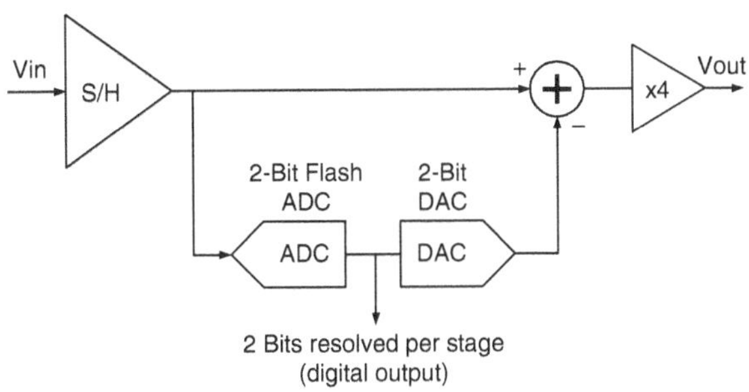

Fig. 3.13 Pipeline stage detail

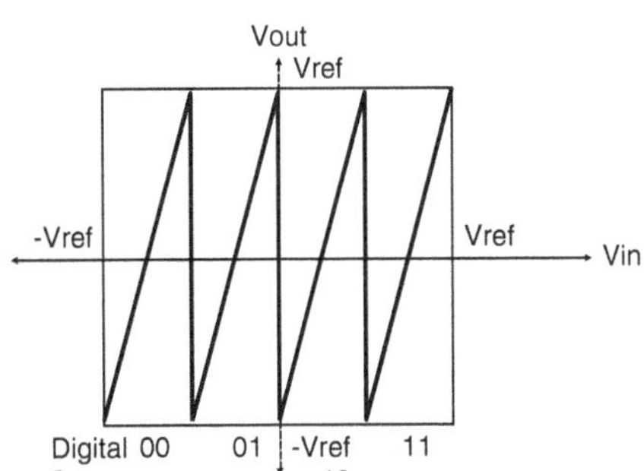

Fig. 3.14 Stage transfer function

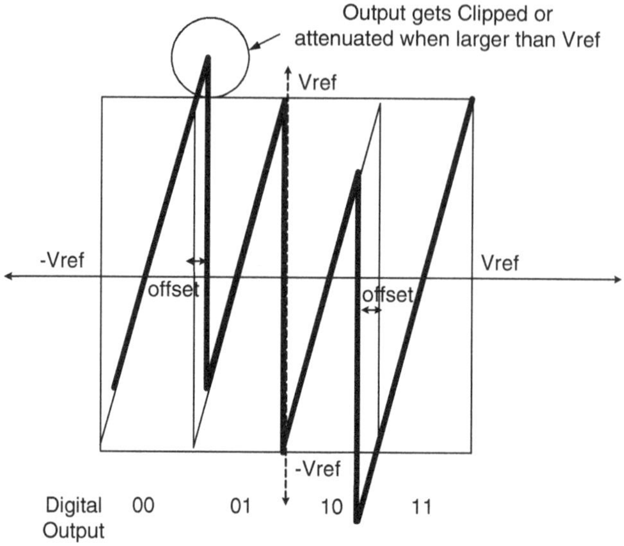

Fig. 3.15 Over-range error with pipeline stage

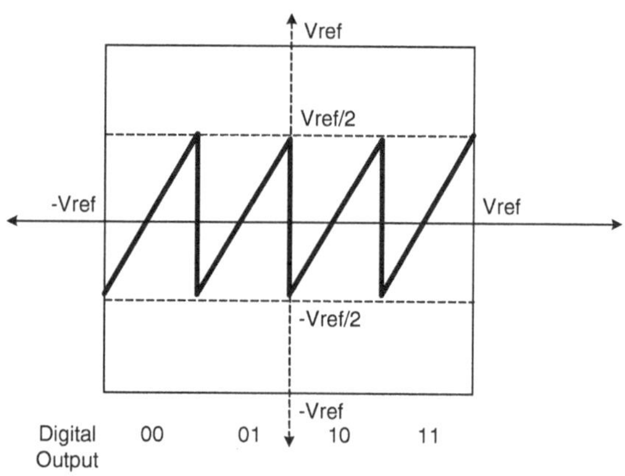

Fig. 3.16 Reduced gain stage transfer function

Hence if the subsequent stage detects an over-range error, the error may be digitally eliminated by adding or subtracting a bit from the digital output (depending on whether the error was an over or under range error). Non-trivial digital subtraction is avoided by altering the transfer function of Fig. 3.16 by adding a Vref/4 offset [18] as shown in Fig. 3.18:

To simplify the sub-ADC, the comparator at Vref can be eliminated, yielding the final transfer function shown in Fig. 3.19.

With three unique digital outputs, the final transfer function is referred to as a 1.5 bit/stage architecture.

Fig. 3.17 Impact of errors on stage transfer function

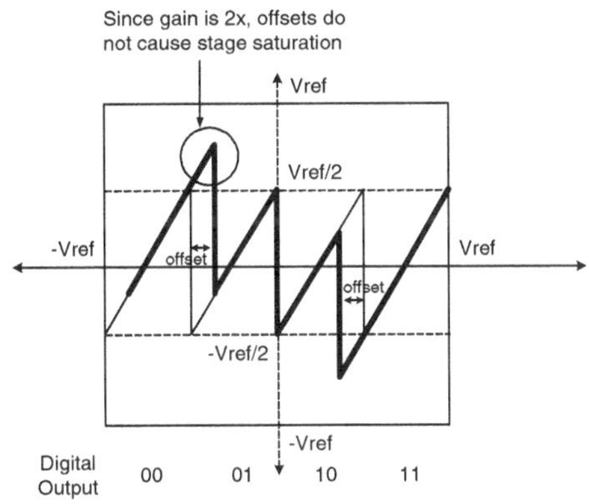

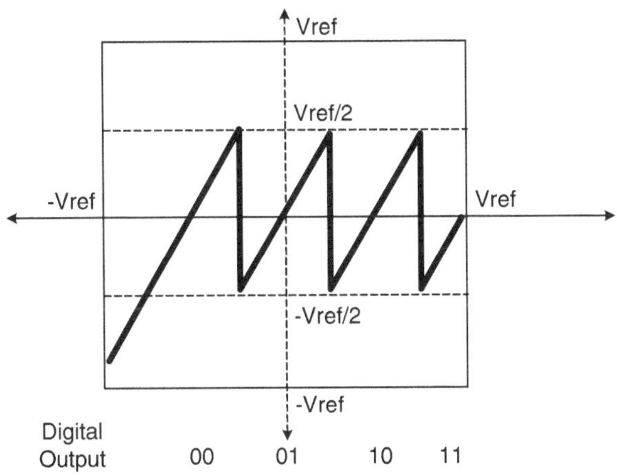

Fig. 3.18 Vref/4 Offset to eliminate digital subtraction

10-bits can be resolved using 1.5 bits/stage with eight such stages, followed by a 2-bit flash stage to resolve the final two bits (error correction cannot be used on the last stage since there is no subsequent stage to correct the error – note the 2-bit flash has thresholds at −Vref/2, 0, +Vref/2). The final 10-bit output code can be realized by digitally combining the outputs from each stage as described in [18]. Figure 3.20 illustrates an example 10-bit pipelined ADC configuration.

3.9 Sub-ADC Design: Comparator

From Section 2.5, it is clear that the number of comparators required in the Flash sub-ADC increase exponentially with each additional bit resolved per stage. As a result, from the standpoint of minimizing complexity in the sub-ADC it is desirable to minimize the number of bits resolved per stage in the sub-ADC.

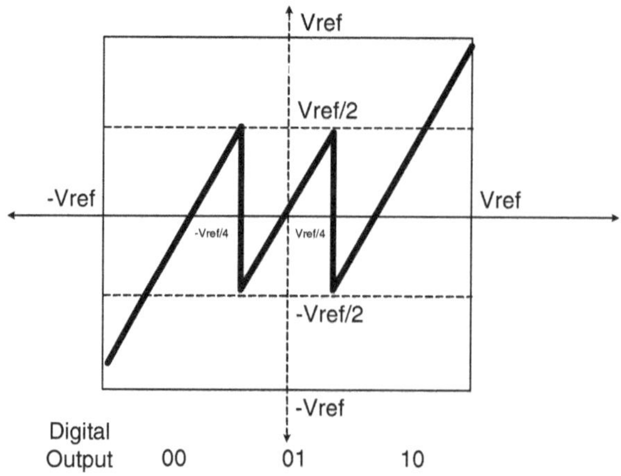

Fig. 3.19 1.5Bit/stage transfer function

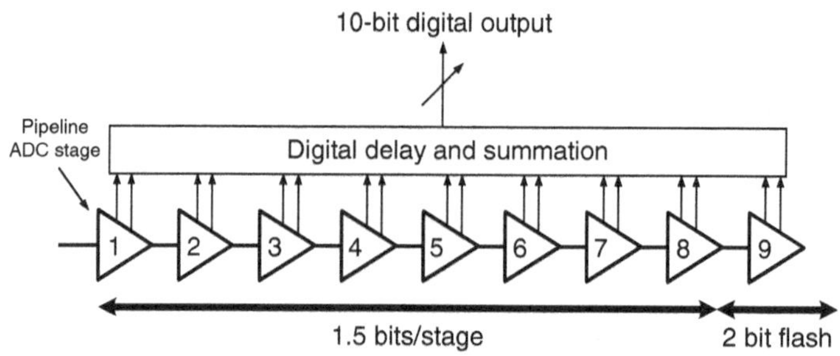

Fig. 3.20 10-Bit pipeline ADC using 1.5 bits/stage

As discussed in Section 3.8, due to redundancy in the pipeline stage, comparators with large offsets can be used in the Flash sub-ADC. Typically dynamic comparators are used for the sub-ADC as they have low power consumption and complexity, but high offset. In [24] different dynamic comparator topologies for use in pipelined ADCs are analyzed.

3.10 Front-End Sample-and-Hold

In a pipelined ADC the analog input is typically sampled by both the MDAC and sub-ADC with different sampling circuits. Due to mismatches in the signal paths as well as threshold mismatches in the sampling switches, the analog input sampled by the sub-ADC and MDAC in the first pipeline stage (when driven by an analog input source) is different, as illustrated in Fig. 3.21.

For a difference in effective sampling time between the sub-ADC and MDAC of Δ_{skew}, and an input sinusoid with frequency f_{in}, and peak voltage V_{peak}, the maximum difference in input voltage sampled by the MDAC and sub-ADC is given by

$$V_{skew-\max} = 2\pi f_{in}\Delta_{skew}V_{peak} \tag{3.21}$$

For very large input frequencies the MDAC and sub-ADC can sample vastly different inputs, resulting in massive harmonic distortion in the ADC output. For example, if a full scale sinusoid of 270 MHz is applied to a pipelined ADC which has a sampling skew of 140 ps between MDAC and sub-ADC, the difference between inputs sampled by the MDAC and sub-ADC can be as high as a quarter of the full scale voltage.

To ensure the sub-ADC and MDAC see the same input, a front-end Sample-and-Hold (S/H) is commonly used before the first pipelined stage so as to make the input to the first pipeline stage discrete time, thus independent of input frequency given a sufficient settling time. In sub-sampled applications where the input frequency to the pipelined ADC can be very high, a front-end S/H ensures the sub-ADC and MDAC in the first stage always see the same input.

Since a front-end S/H has a gain of one, its inherent thermal noise contributes directly to reduce the dynamic range of the ADC. To thus maintain input dynamic range, the S/H is required to have a noise floor and distortion lower than that of the pipelined ADC following it. As a result the power of the ADC is significantly increased by using a front-end S/H. It is not unusual for the front-end S/H to be the largest power consumer in an ADC. Figure 3.22 illustrates a common topology used to implement a front-end S/H [25].

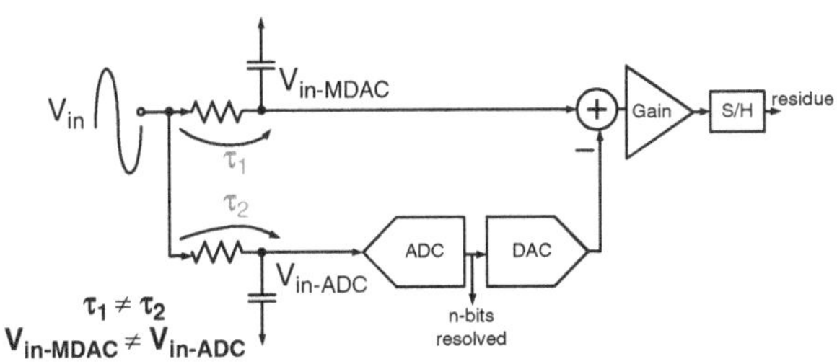

Fig. 3.21 Timing mismatch between sub-ADC and MDAC when first pipeline stage is connected directly to analog input

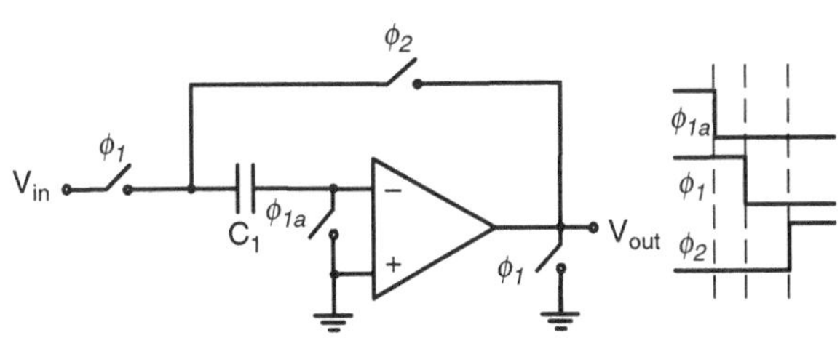

Fig. 3.22 Commonly used, 'flip-around' front end S/H topology

3.11 Summary

In this chapter a detailed discussion of pipelined ADCs was given. Circuit issues and tradeoffs discussed included: opamp DC gain, opamp bandwidth, thermal noise, capacitor matching, and sub-ADC comparator design. The role of a front-end sample-and-hold was also discussed, where it was shown that a front-end sample-and-hold ensures functionality for very high input frequencies.

Chapter 4
Scaling Power with Sampling Rate in an ADC

4.1 Overview

This chapter discusses design issues related to a reconfigurable ADC which has its power scalable with sampling rate. A comparison of digital versus analog power is given, where current scaling is shown as an analog power scaling technique. The consequences of extended power scaling using current scaling are analyzed where excessive current scaling is shown to increase design and simulation difficulty, resulting in poorer yield due to larger mismatches, increased bias voltage sensitivity, and IR drops. The goal of this chapter is to provide the reader with an appreciation of the limitations of conventional power scaling techniques, and thus provide the motivation for the state of the art power scaling technique used in Chapters 7 and 8.

4.2 ADC Power as a Function of Sampling Rate

From (2.2), the ADC figure of merit is a function of the ratio of power to sampling rate:

$$FOM = \frac{Power}{(2^{ENOB})(f_s)}$$

Thus to maintain a fixed figure of merit as the sampling rate is decreased, the power must also decrease. Designing a pipelined ADC with very low power for low sampling rates however is a non-trivial task. Unlike digital power, analog power does not automatically scale with sampling rate, since analog power is given by the product of supply voltage and average current consumed – where in analog circuits neither voltage nor current explicitly scale with the operating speed. Thus a power scalable ADC requires techniques to make the analog power an explicit function of sampling rate.

I. Ahmed, *Pipelined ADC Design and Enhancement Techniques*, Analog Circuits and Signal Processing,
DOI 10.1007/978-90-481-8652-5_4,

4.3 Digital Versus Analog Power

Digital circuits primarily operate transistors in triode and cut-off regimes, whereas in analog circuits transistors primarily operate in the saturation regime. Steady state outputs in digital circuits are realized by charging a load capacitance through a triode switch to a supply voltage as shown in Fig. 4.1.

Thus digital circuits only require enough power to charge/discharge the load capacitance to the final logic level. For a full cycle from zero to one then back zero $Q = CV_{DD}$ is transferred from V_{DD} to ground in Fig. 4.1

$$\therefore E_{cycle} = QV_{DD} = CV_{DD}^2 \tag{4.1}$$

$$\text{since } P = \frac{E}{T} = Ef \tag{4.2}$$

$$\therefore P = CV_{VDD}^2 f \text{ [26]} \tag{4.3}$$

Thus, assuming the digital circuitry in an ADC is clocked by the sampling clock, the average digital power automatically scales with sampling frequency. In most ADCs however, the digital power consumes only a small fraction of the total power, thus if ADC power is to scale with sampling frequency, analog power must also scale with sampling frequency.

As analog circuits require static bias currents to bias transistors in the active region, analog power is given by the product of two static quantities: $P_{ana\log} = IV$. Thus if analog power is to scale with sampling frequency, voltage and/or current must be made functions of the sampling frequency (i.e.: $P(f_s) = i(f_s)V(f_s)$). Power scaling by supply voltage scaling is not a viable option as reducing the supply voltage reduces signal swing, possibly moving saturated devices into the triode region, and/or significantly reducing the ADC SNR due to reduced signal swings. As minimum signal swings are required in analog circuits, power scaling by voltage reduction can only provide a minimal power-speed dependency. Analog power scaling is commonly achieved by making the bias currents a function of sampling frequency [27, 28]. In [27], total ADC power is scalable between 3 and 220 MS/s,

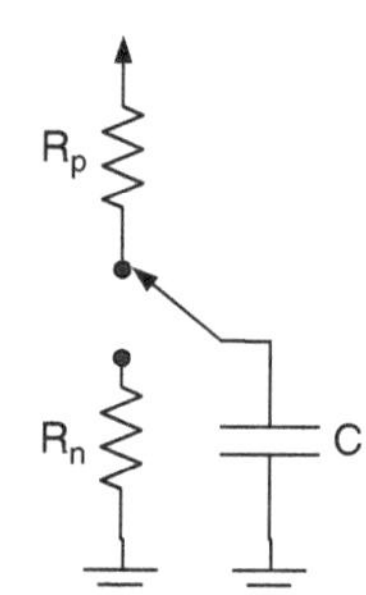

Fig. 4.1 RC model of digital switching

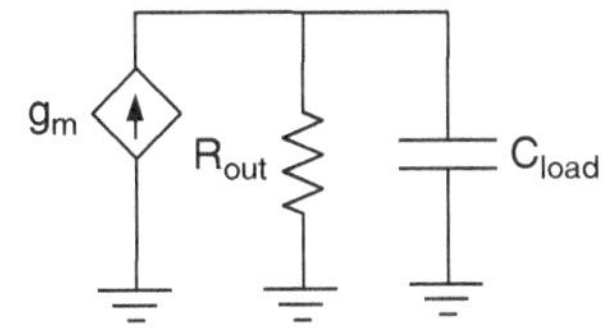

Fig. 4.2 Simplified small signal opamp model

and in [28] opamp bias currents are shown to be scalable between effective sampling rates of 10 kS/s–10 MS/s; both implementations use current scaling to reduce bias currents with operating speed. Assuming a first order response of an opamp as shown in Fig. 4.1: RC model of digital switching

Figure 4.2, opamp unity gain frequency is given by

$$\omega_{ta} = \frac{g_m}{C_{load}} \tag{4.4}$$

$$\therefore \omega_{ta} = \frac{1}{C_{load}} \sqrt{2\mu C_{ox} \frac{W}{L} I_D} \tag{4.5}$$

Thus reduction of bias currents with sampling frequency reduces the bandwidth of the opamp, which is consistent with the concept of maintaining a constant figure of merit, i.e. since the sampling frequency is reduced, the opamps do not require as high a bandwidth as more settling time is available.

According to square law equations, as transistor drain–source currents are reduced, transistor $V_{GS} \rightarrow V_t$

$$\text{since } V_{GS} = \sqrt{\frac{2I_D}{k^W/_L}} + V_t, \tag{4.6}$$

$$\therefore \lim_{I_D \to 0} V_{GS} = V_t \tag{4.7}$$

However as transistor V_{GS} tends to V_t, the channel region below the gate oxide becomes less inverted [29] (referred to as weak inversion), such that the inversion channel bridging the source and drain becomes diffusion carrier dominated, rather than drift carrier dominated as is the case in strong inversion. Thus like BJTs (which have a current dominated by diffusion), MOS transistors for low bias currents have a current that is exponentially related to the gate–source voltage [30]. Hence the I_{DS}–V_{GS} relation deviates from square law when $V_{GS} \rightarrow V_t$.

An advantage of weak inversion operation is due to the exponential dependency of current on gate–source voltage, the g_m/I_D ratio (i.e. transistor gain) is a maximum in weak inversion [31]. Weak inversion operation is commonly used in analog circuits that require very low power consumption. A significant disadvantage of operation in the weak inversion region however is the lack of continuous, easy to manipulate models of transistor operation in weak inversion. As such design in weak inversion is often avoided where power requirements are not stringent, since careful design would require much background knowledge in weak inversion operation as well as patience to deal with complicated device parameters [32]. Furthermore as most ADCs are designed in digital or logic processes (which rarely operate transistors in weak inversion), transistor simulation models may not be well characterized in the weak inversion regime. Thus a functional ADC that relies on poor weak inversion models could take several fabrication iterations before all desired specifications are met [32].

4.4 Weak Inversion Model: EKV

This section briefly discusses a popular model (EKV) [32], which describes transistor operation in both strong and weak inversion regions. In the EKV transistor model, drain-source current is given as the difference between a forward current, and a reverse current [32]:

$$I_{DS} = I_F - I_R \tag{4.8}$$

where the forward current depends on gate and source voltages, and the reverse current depends on gate and drain voltages. For an NMOS transistor the current components can be expressed as [32]

$$I_{F(R)} = \frac{W}{L} I_s \log^2\left(1 + e^{\frac{\kappa(V_G - V_{T0}) - V_{S(D)}}{2U_T}}\right) \tag{4.9}$$

where $\kappa \equiv \frac{C_{ox}}{C_{ox}+C_{dep}}$ is the reciprocal of the sub-threshold slope factor, V_{T0} is the zero-bias threshold voltage, and $U_T \equiv \frac{kT}{q}$ is the thermal voltage. I_s is the specific current which is roughly twice the threshold current of a square transistor (note I_s is NOT the source current) and is given as

$$I_s = \frac{2\mu C_{ox} U_T^2}{\kappa} \tag{4.10}$$

If the forward current is much larger than the reverse current, the channel current depends on V_{GS}, and becomes largely independent of the drain potential – hence the transistor is saturated. If I_F is comparable to I_R however, the channel current depends on drain and source potentials, hence the transistor is in the ohmic or triode region.

The *inversion coefficient* (IC) describes the level of channel inversion, and is given as

$$IC = \frac{I_D}{I_s} \tag{4.11}$$

A transistor is in strong inversion if IC > 10, moderate inversion if IC $\sim$ 1, and in weak inversion if IC < 0.1 [33]. In terms of V_{eff}, this typically translates into strong inversion for V_{eff} > 220 mV, weak inversion for $V_{eff} < -72$ mV, and moderate inversion between weak and strong inversion, i.e. $V_{eff} \approx 40$ mV [33].

4.5 Weak Inversion Issues: Mismatch

A major disadvantage of transistor operation in weak inversion is an increased current mismatch. The current mismatch of two transistors in weak inversion that have the same V_{GS} (e.g. a current mirror) is given by

$$\sigma_{I_D} = \sqrt{\sigma_\beta^2 + \left(\frac{g_{mg}}{I_D}\sigma_{Vt}\right)^2} \tag{4.12}$$

where $g_{mg} = \frac{I_D}{nU_T}\sqrt{\frac{1}{i_{f1}+\frac{1}{2}\sqrt{i_{f1}+1}}}$, $i_{f1} = \frac{I_D}{2n\beta U_T^2}$, and n is a fitting factor between 1 and 2 [32]. The relationship between current mirror mismatch, area, and bias current is illustrated in Fig. 4.3.

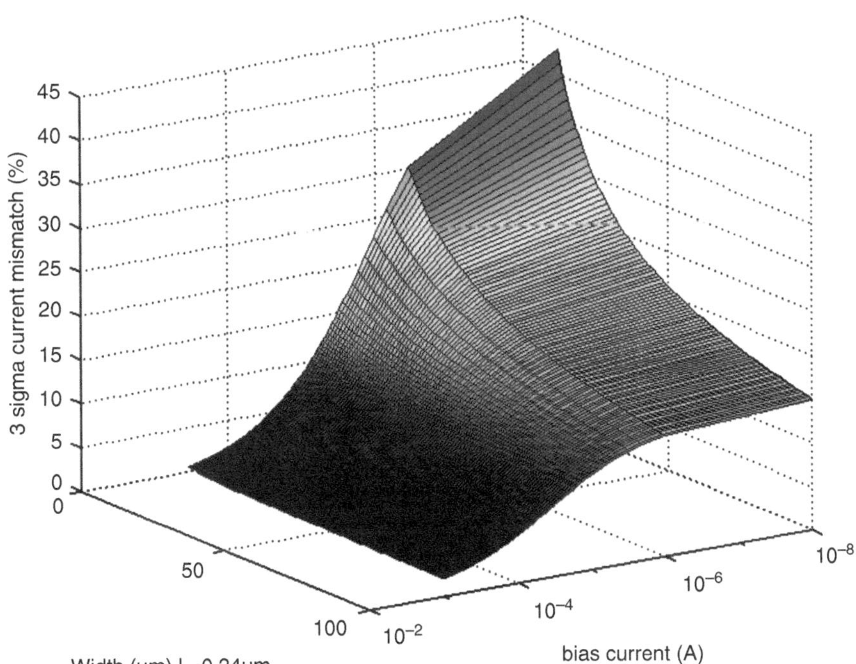

Fig. 4.3 3σ Current mismatch versus device area and bias current

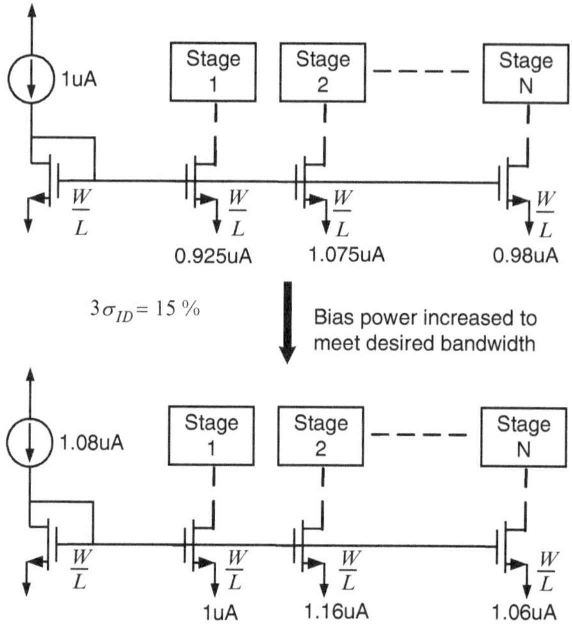

Fig. 4.4 Illustration of impact of mismatched current sources

From Fig. 4.3 it is clear as the bias current decreases (placing the transistor deeper into weak inversion), the 3σ mismatch of the mirror current increases significantly [32]. One consequence of the current mismatch is a sub-optimal distribution of power in an ADC. For example, consider Fig. 4.4, where the opamps of several stages in a pipeline ADC are biased with a single current mirror, which has $3\sigma = 15\%$ mismatch in current (i.e. $\pm 7.5\%$ peak variation). Since a pipeline is limited by the slowest stage, potentially the power could have to be increased by over 15% to meet the desired bandwidth.

Clearly a 15% + increase in power is not desirable – especially since much of the excess power is wasted. Often to avoid the high mismatch in current mirrors, mirror transistors are designed with a large area, but small W/L ratio so as to maintain strong inversion (e.g. $V_{eff} \sim 400$ mV). Such an approach however requires a large area overhead to maintain strong inversion over large variations in bias current. (For example) in [34] to maintain a current mirror transistor in strong inversion for a bias current of 25 nA in a 0.35 μm process, a W/L ratio of 3/50 μm is used for the current mirror transistors. Note that a current source transistor sized at 3/50 μm cannot be used for higher bias currents without V_{eff} becoming prohibitively large. Thus if the current source transistor were used to bias an opamp for various current scaled values (for different sampling rates), an array of different current mirrors must be used (to maintain strong inversion for different bias currents), thus consuming additional area.

For deep submicron technologies leakage current can become a significant issue if bias currents are on the order of the leakage current (which increases with transistor size). With bias currents on the order of leakage currents, reliable analog circuit design can become difficult, as transistors cannot be accurately biased with desired drain-source currents. In other words, one cannot necessarily guarantee if a device is in active or triode.

4.6 Current Scaling: Multiple Design Corners

Although simple to implement, current scaling has the disadvantage that it necessarily increases the number of design corners in the ADC. As the range of bias currents between minima and maxima are required to be verified over temperature and process corners, design/simulation time for an ADC using current scaling as a power scalable technique can be excessive (and thus expensive). Multiple bias currents also increase post fabrication test time, hence increasing cost, as the ADC must be verified at all corners to ensure a working product is delivered to a potential customer.

4.7 Current Scaling: Bias Point Sensitivity

Consider the differential ${}^{dI}/_{dV_{gs}}$ for a transistor in strong and weak inversion:

Current sensitivity (Strong Inversion)

$$\frac{dI}{dV_{GS}} \approx 2k(V_{GS} - V_t) \tag{4.13}$$

Current sensitivity (Weak Inversion)

$$\frac{dI}{dV_{GS}} \approx e^{\frac{V_G - V_t - V_s}{nU_T}} \tag{4.14}$$

In strong inversion the rate of change of drain-source current varies linearly with V_{GS} variation, whereas it is exponential in weak inversion. Thus if a transistor is acting as a current source to an opamp is in weak inversion, a small variation of gate–source voltage on the transistor due to (e.g.) noise coupling from a nearby digital circuit, thermal fluctuations of a resistor acting as a reference current source, or threshold mismatch, will cause the unity gain frequency of the opamp, hence accuracy of the ADC to fluctuate significantly. An analogous problem manifests with the biasing of BJTs (which have an exponential relation between current and base–emitter voltage). To reduce current sensitivity, BJTs use emitter degeneration to reduce the transistor gain, at the cost of reduced signal swing and increased power. Although degeneration resistors could be switched in and out in a power scalable ADC to reduce mismatch for different bias currents, multiple design corners still remain, thus a significant amount of time would be required to verify the ADC overall design corners.

Thus although it is possible to operate an opamp (hence a pipeline ADC) while deep in the weak inversion region, the high sensitivity of the bias nodes makes current scaling over large ranges (such that transistors are driven deep into weak inversion) an impractical approach to achieve lower power for low f_s. The high sensitivity to bias fluctuations could be significant if the ADC were part of a larger, noisier digital system, where fluctuations of a few millivolt could easily be induced on opamp bias nodes from (e.g.) substrate noise.

4.8 Current Scaling: IR Drops

In some cases an IR drop is required across a resistive load to provide a specific gain. If bias currents are reduced, the gain and signal swing also reduce since:

$$A \approx g_m R = R\sqrt{2\mu C_{ox}\frac{W}{L}I_D} \tag{4.15}$$

therefore as current decreases, gain decreases. An example of resistor dependent gain is shown in Fig. 4.5, specifically a pre-amp which is commonly used in Flash ADC comparators, and multi-bit/stage pipeline ADCs comparators. Alternatively active loads can be used to provide gain, as shown in Fig. 4.6. Such configurations have a gain which increases with reduced bias currents since:

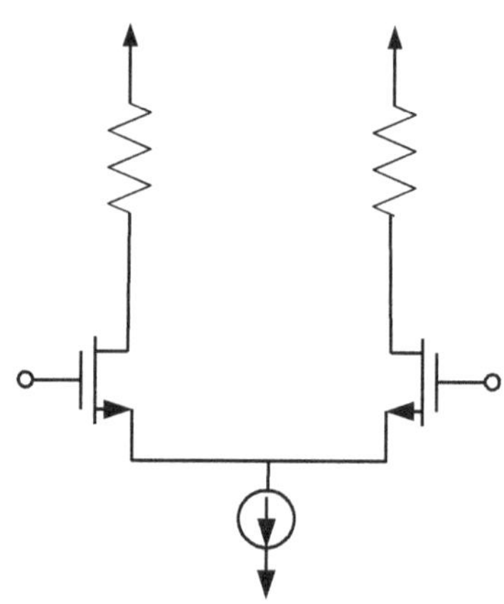

Fig. 4.5 Differential pair with RC load

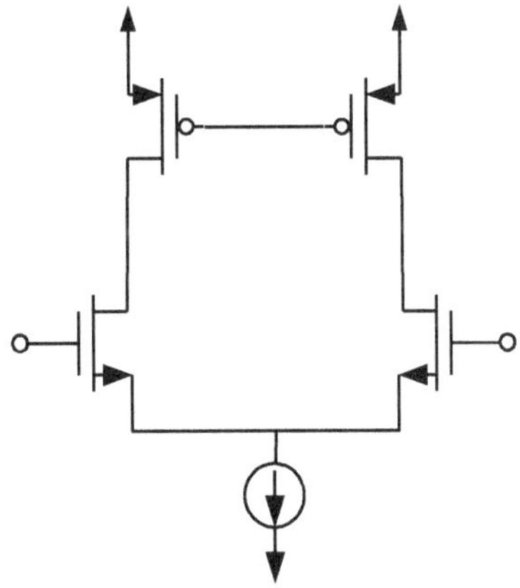

Fig. 4.6 Differential pair with active load

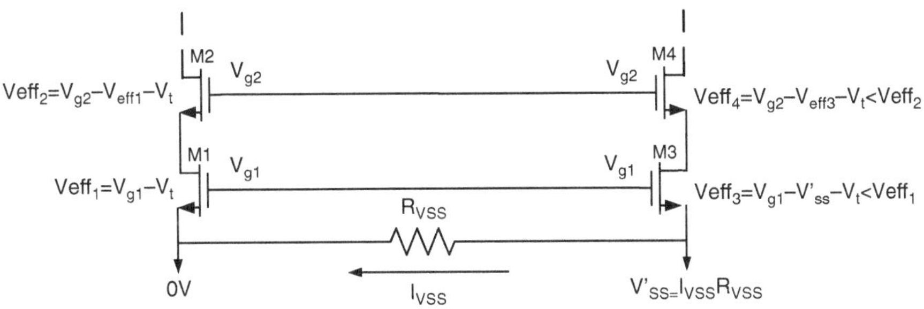

Fig. 4.7 Impact of low currents on IR drops

$$A \approx g_m r_{ds} = \frac{\sqrt{2\mu C_{ox}\frac{W}{L}I_D}}{\lambda I_D} = \frac{1}{\lambda}\sqrt{\frac{2\mu C_{ox}\frac{W}{L}}{I_D}} \tag{4.16}$$

therefore as current decreases, gain increases. Active differential loads however require common-mode feedback to have a defined output common-mode.

IR drops across power supplies also pose a potentially significant design issue for weak inversion operation. As mentioned in Section 4.7, current mirror transistors have a high sensitivity to bias node fluctuations, thus it is possible that even a small IR drop of a few millivolt between mirror transistor supply voltages (due to e. g. physical separation on a larger chip) could cause significant current mismatch hence potentially reduced performance (e.g. Fig. 4.7).

If the transistors of Fig. 4.7 however are in strong inversion, the current is a much weaker function of V_{GS}, thus small V_{GS} variations due to IR drops have a much smaller impact on the desired bias current.

4.9 Summary

This chapter discussed the dependency of power with sampling rate for analog and digital systems. Current scaling was shown as common technique to reduce analog power with sampling rate. It was shown that current scaling drives MOS transistors deep into the weak inversion region for extended reductions in sampling rate, where due to less accurate models, circuit design/fabrication could take several iterations to meet desired performance. Increased mismatch, bias point sensitivity, and IR drops were also shown as limiting factors to the extent to which current scaling can be used to scale analog power with operating speed.

Chapter 5
State of the Art Pipelined ADC Design

5.1 Overview

In this chapter state of the art research in pipelined ADC design techniques which address some of the circuit issues and tradeoffs discussed in Chapters Two through Four will be discussed. The issues most relevant to the silicon implementations presented in Section II of this book, namely rapid calibration, reconfigurable design, and low power design will be discussed in most detail. In Section 5.2 the concept of calibration in pipelined ADCs is introduced as a way of reducing the impact of non-idealities such as low DC opamp gain, and capacitor mismatch. Examples of pipelined ADCs which use calibration to improve their performance will be discussed, where the tradeoffs and limitation in each approach will be discussed. Section 5.3 presents techniques which can be used to achieve a power which scales with sampling rate in pipelined ADCs. In Section 5.4 techniques to reduce power consumption in pipelined ADCs will be reviewed. The goal of this chapter is to gain an appreciation of state of the art solutions and their limitations to many of the design issues discussed in earlier chapters, and thus motivate the solutions presented in Section II of this book.

5.2 Calibration in Pipelined ADCs

As discussed in Sections 3.4 and 3.7, low DC opamp gain and capacitor mismatch are major sources of missing codes which degrade the linearity of a pipelined ADC. To enable higher linearity in processes which offer only low opamp DC gain and/or insufficiently matched capacitors, digital calibration techniques which measure and compensate the effect of missing codes can be used. As newer deep sub-micron technologies offer digital circuits with lower power consumption and area compared to older technologies, digital calibration has become increasingly attractive to realize low power highly-linear ADCs. Sections 5.2.2 and 5.2.3 review how gain

I. Ahmed, *Pipelined ADC Design and Enhancement Techniques*,
Analog Circuits and Signal Processing,
DOI 10.1007/978-90-481-8652-5_5, © Springer Science+Business Media B.V. 2010

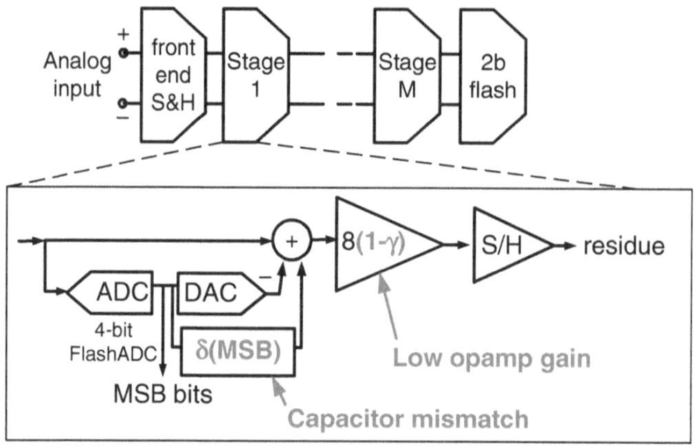

Fig. 5.1 Illustration of errors sources in pipelined ADC

Fig. 5.2 Ideal residue transfer curve of a 3 + 1-bit pipelined stage

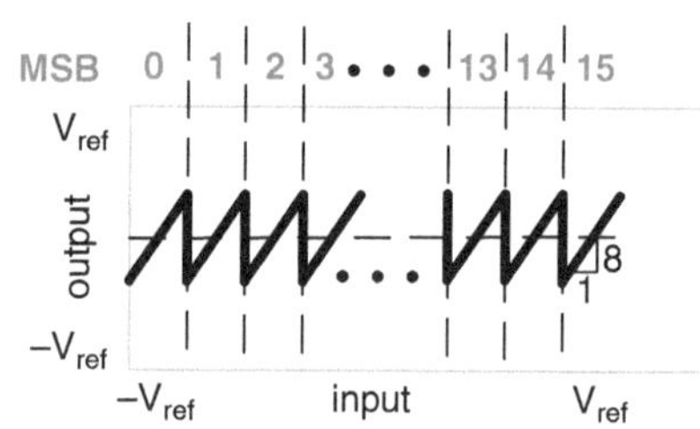

and DAC errors are corrected in a pipeline stage. Sections 5.2.4 and 5.2.5 compare and contrast foreground and background calibration techniques.

5.2.1 Review of Error Sources

Figure 5.1 illustrates a 3 + 1-bit pipelined ADC stage which has its gain altered from the ideal value by $(1 - \gamma)$, and DAC output in error by $\delta(\mathrm{MSB})$.

Figures 5.2–5.4 illustrate the residue transfer curve of Fig. 5.1 when the pipelined stage is: (1) ideal, (2) has a non-ideal gain, and (3) has a non-ideal gain *and* DAC errors respectively.

5.2.2 Gain Error Correction

Gain errors can be corrected by simply scaling the digital backend ADC output by the inverse of the gain error factor $(1 - \gamma)$. Figure 5.5 illustrates an example of an

Fig. 5.3 Residue transfer curve of a 3 + 1-bit pipelined stage with gain errors included

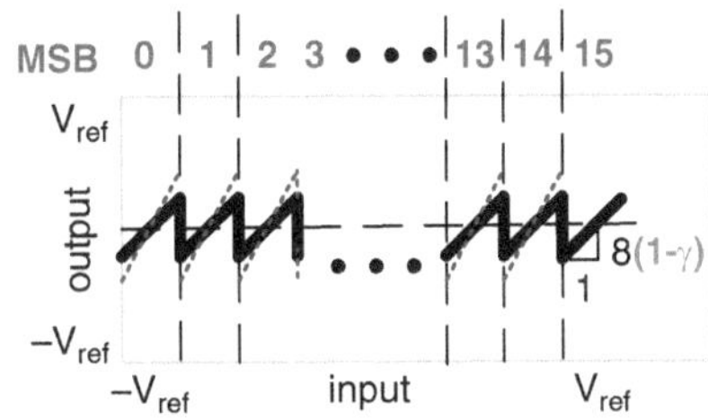

Fig. 5.4 Residue transfer curve of a 3 + 1-bit pipelined stage with gain and DAC errors included

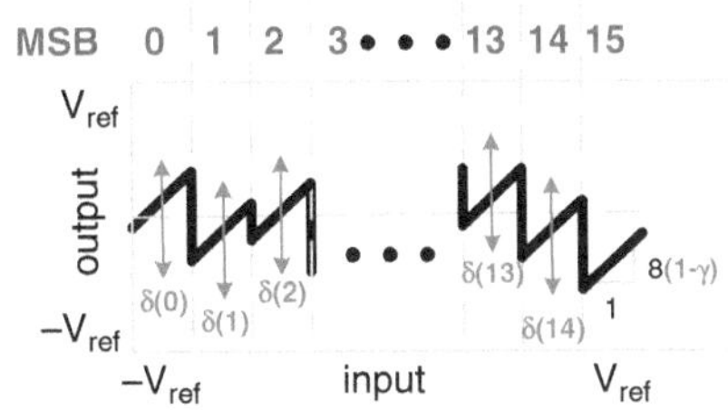

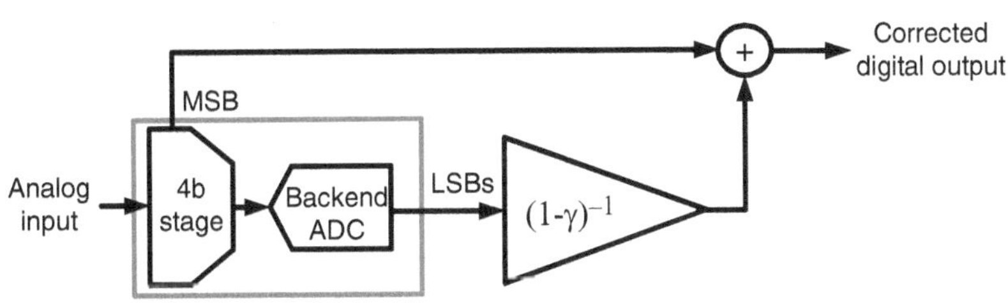

Fig. 5.5 Gain error correction of first pipeline stage

architecture which compensates for the effect of the non-ideality γ in the first pipeline stage, assuming the value of γ is already known. Gain errors in subsequent pipeline stages can be corrected by inductively using the same error correction technique.

Most stages which resolve 1.5-bits/stage use the circuit topology shown in Fig. 5.6 for the MDAC.

From the topology it is noted that the DAC is implemented with only one capacitor (C_1), thus there is no DAC error in such topologies. As the majority of pipelined ADCs in literature use 1.5-bit/stage topologies, most calibration techniques in literature deal only with the correction of γ. For stages which resolve more than 1.5-bit/stage however, the DAC error can be significant.

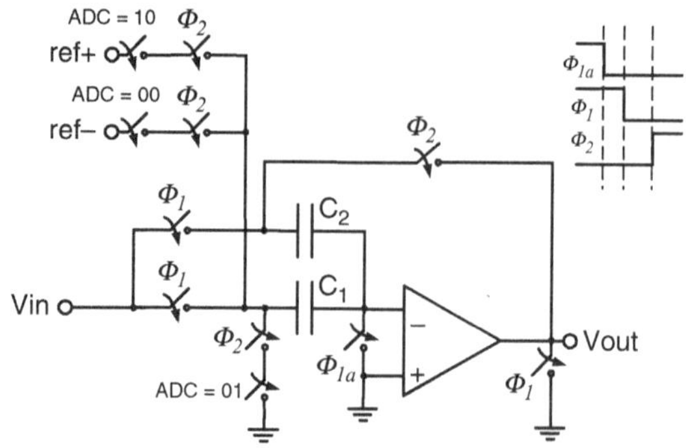

Fig. 5.6 Commonly used circuit topology for 1.5-bit based MDAC

5.2.3 DAC Error Correction

From Section 3.7 it was shown that capacitor mismatch in the DAC results in unique missing codes at every MSB transition, yielding substantial harmonic distortion. Since unique errors are produced at each MSB transition a stage-gain calibration scheme (which effectively only corrects for the average number of missing codes at each transition) cannot be used to digitally correct the ADC output. To correct DAC errors (e.g. [44–46]) a separate corrective term for *each* MSB transition is required, significantly increasing the complexity of the correction scheme over gain-only correction techniques. For example, with a 3 + 1-bit pipeline stage, 15 correction parameters for 16 unique DAC outputs are required to be estimated, whereas a stage-gain correction scheme has only one parameter to estimate.

Comparing Figs. 5.3 and 5.4, it is noted that missing codes produced by gain errors look the same as missing codes produced by DAC errors where the DAC error is constant at every MSB transition. Thus in a DAC calibration scheme (where the missing codes are corrected as a function of *each* MSB), the gain errors are also corrected in addition to DAC errors. Figure 5.7 illustrates an example of a system which corrects for both DAC and gain errors in the first stage of a pipelined ADC, when the non-idealities δ(i) are already known.

5.2.4 Foreground Calibration

Sections 5.2.2 and 5.2.3 discussed how gain and DAC errors can be corrected when the amount of error is already known in advance. In reality however the error is unknown to the designer before fabrication; furthermore the magnitude of each error source varies from chip to chip due to process variation. Thus a scheme to adaptively measure the unknown and unique error sources in an ADC needs to be implemented. In Fig. 5.8, a foreground calibration scheme is shown.

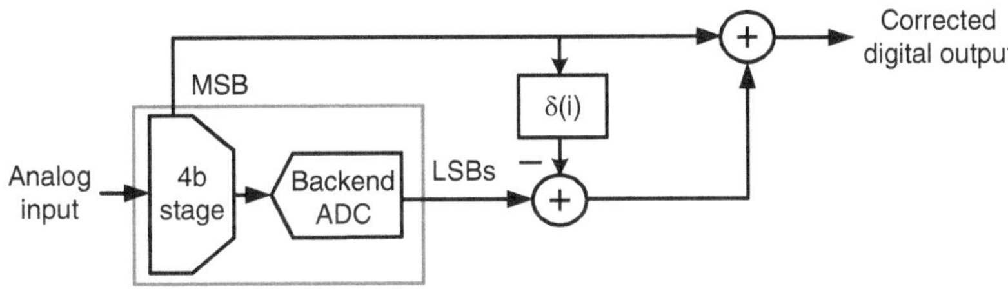

Fig. 5.7 Correction of gain and DAC errors in first pipeline stage

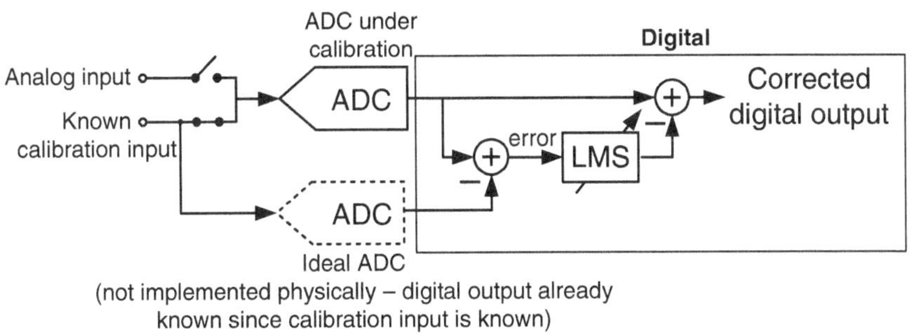

Fig. 5.8 Principle of foreground calibration

Foreground calibration estimates the unknown errors sources by interrupting normal ADC operation and applying a known input sequence to the ADC. By comparing the output of the ADC to the expected ADC output under ideal conditions (i.e. no non-idealities) the impact of each error source can be measured and corrected. Examples of foreground calibration in publications can be found in [47, 48].

The advantage of a foreground scheme is that calibration can be achieved within a small number of clock cycles, since the error signal labeled in Fig. 5.8 is highly correlated with the error sources causing the missing codes. The disadvantage of foreground calibration is that the ADC is required to be taken offline every time calibration is performed, which in some applications may not be possible.

5.2.5 Background Calibration

Background calibration continuously measures and corrects the effect of non-idealities in a pipeline stage, thus has the significant advantage that the ADC is not required to be taken offline to perform calibration. As such the vast majority of calibration based publications are focused on background techniques.

Several topologies have been proposed recently to implement background calibration, where the vast majority of the schemes use a statistics based approach to

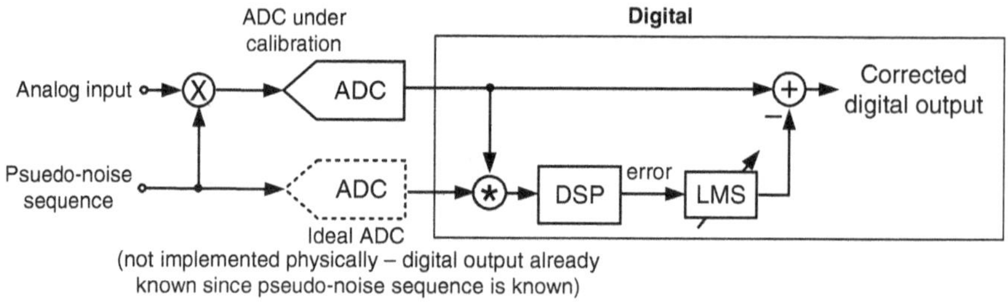

Fig. 5.9 Principle of background calibration

realize calibration. In a statistical scheme, the input of the pipeline stage under calibration is combined with a known pseudo random sequence, where by correlating the digital output of the ADC with the known pseudo random sequence, the impact of missing codes can be determined. To avoid significantly altering the ADC output spectrum, the pseudo-noise sequence is typically made very long to avoid correlations with the analog input, as well as small in amplitude so that the injected pseudo random sequence which appears as an additional white noise source at the output only consumes a small portion of the dynamic range. Figure 5.9 shows the basic principle of background calibration.

With statistics based background calibration schemes however, since the digital output of the ADC is highly correlated with the analog input and weakly correlated with the pseudo random sequence, a large number of clock cycles are required to accurately extract the pseudo random sequence from the digitized analog input in the ADC output. For example, in [46] $\sim 10^7$ cycles were required to achieve 13-b linearity, and in [45] $\sim 10^8$ clock cycles were required to achieve >14b linearity. In [49] it was shown empirically that statistical techniques required on the order of 2^{2N} clock cycles to calibrate gain errors only. For 11-bit linearity approximately four million clock cycles are required to only correct gain errors using statistics-based background calibration. Thus while background schemes are popular as they enable continuous ADC operation, the calibration time of background approaches is very lengthy. Example publications which use background calibration are: [50–60]

It is noted that in DAC calibration schemes, all MSB bits are required to be exercised to measure each unique DAC error (as the DAC error is different for each MSB). Since an MSB bit can change at most only once every clock cycle, DAC calibration schemes can require $M - 1$ times more cycles to converge than gain calibration schemes (where M is the number of unique MSB outputs).

5.2.6 *Rapid Calibration of ADC Errors*

In an industrial environment where ICs are mass produced, ICs are tested for functionality by automated testers. In ADCs which use background-statistical techniques to achieve calibration, long calibration times can lead to excessive

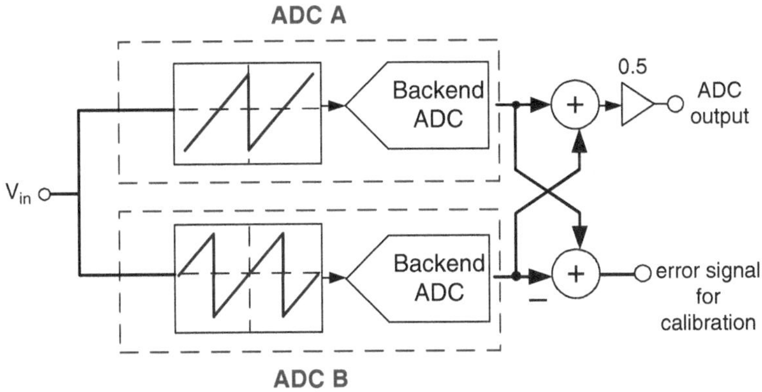

Fig. 5.10 Split-ADC topology

test times thus limiting IC production throughput and hence revenue. For example, with four million calibration cycles, even with a reasonably high sampling rate of 40 MS/s, one tenth of a second would be required at minimum to test each ADC. For higher resolution and/or lower speed ADCs the test time can be much higher [49]. In the interest of larger production throughput it is highly desirable to reduce calibration time.

Reducing calibration time has become an active area of research over the past few years. One topology which has proven to be highly effective in reducing calibration times in background schemes is the 'dual-ADC' or 'split-ADC' approach [49, 54, 58]. As shown in Fig. 5.10, the split-ADC takes a single ADC and splits it into two almost identical ADCs where each ADC has half the area, and half the thermal noise floor (thus half the power) of the overall ADC. The final ADC output is derived by taking the average of each ADC output, hence power and area of the split-ADC topology to a first order are not increased over a conventional ADC [49].

Each ADC is identical, except the residue transfer curve of the stage under calibration in one ADC is designed differently than the other. As a result when the ADCs are free of errors both ADCs produce the same output, however when errors are included each ADC produces different outputs. Since the analog input effectively appears as common mode to the split-ADCs, the error signal which is formed by the difference of the two ADCs is very weakly correlated to the analog input. However the error sources are very highly correlated with the difference in ADC outputs (i.e. error signal) due to each residue transfer curve being designed slightly differently than the other signal [54]. Thus error sources can be estimated very quickly in the background by only looking at a small number of clock cycles of the error signal.

It should be noted that the split-ADC technique while very effective at reducing calibration time, has thus far only been demonstrated to correct gain errors in 1.5 b/stage based ADCs. In multi-bit pipelined stages (i.e. 2 bits resolved per stage), the effect of DAC errors also needs to be measured and corrected. To date all DAC background calibration techniques in literature use statistical techniques to achieve calibration, and hence suffer from very lengthy calibration times. As noted in Section 5.2.3, since DAC errors are a function of the MSB of a pipeline stage, DAC error measurement can take much longer than gain error calibration, hence in an industrial environment it is even more critical for calibration schemes which correct DAC errors to have short calibration times.

5.3 Power Scalability with Respect to Sampling Rate

The growing demand for mobile systems which can provide multi-standard compatibility in a single solution [1] has stimulated much research in systems which allow for multiple design specifications to be met with only one low power design (e.g. [28–68]). A single reconfigurable solution significantly reduces design time (and thus time to market). In Chapter 4 a review of current scaling was given which was shown to be the most commonly used approach to achieve scalable power with sampling rate in prior ADCs (e.g. [27, 28, 69]). There are also commercial ADCs which have a scalable power. High-speed commercial architectures ($f_{s\text{-}max} > 10$ MS/s) achieve lower power for lower sampling rates using current scaling, and are shown in datasheets to have a small power scalable range (<1:100), likely due to poorer yield at lower sampling rates as transistors are driven into weak inversion. Low-speed architectures (<500 kS/s) achieve power scalability by powering off the ADC between conversion samples. Due to the slow power on/off times of ADCs however, the technique of powering off an ADC between conversions is limited to slower architectures. A survey of power scalable ADCs in industry is given in Table 5.1.

5.4 Power Reduction Techniques in Pipelined ADCs

In many electronic systems achieving low power consumption is a key performance metric. In mobile systems low power consumption allows for increased battery life, and in wired systems low power consumption can in some cases enable low cost packaging – in both cases allowing for more cost effective and thus attractive system solutions. As a result a large percentage of papers published on pipelined ADCs discuss techniques to reduce power consumption. In this section the current state of the art in power reduction techniques in the areas most relevant to this book will be discussed: the elimination of the front-end S/H, and the use of non-opamp based amplifiers to reduce power.

5.4.1 Front-End S/H Removal

As discussed in Section 3.10, a front-end S/H is commonly used to ensure the MDAC and sub-ADC in the first pipeline stage see the same input. However as was

Table 5.1 Survey of power scalable ADCs in industry

Power versus sampling speeds for various ADCs						
Power scaleable ADCs in industry						
Company	Model	Speed (Msps)	Power (mW)	SNDR	ENOB	P. Scaling method[a]
ADI	7467	0.05	0.48	61	9.84	A
ADI	7467	0.1	0.62	61	9.84	A
ADI	7811	0.001	0.0315	58	9.34	A
ADI	7811	0.01	0.315	58	9.34	A
ADI	7811	0.1	3.15	58	9.34	A
ADI	7811	0.35	10.5	58	9.34	A
ADI	9203	5	17	60	9.67	B
ADI	9203	40	75	59.3	9.56	B
Maxim	Max1086	0.001	0.00405	61	9.84	A
Maxim	Max1086	0.01	0.0351	61	9.84	A
Maxim	Max1086	0.05	0.1755	61	9.84	A
Maxim	Max1086	0.1	0.351	61	9.84	A
Maxim	Max1086	0.15	0.54	61	9.84	A
Nordic VLSI	nAD1050-18	10	8	59	9.51	B
Nordic VLSI	nAD1050-18	50	33	59	9.51	B
Fairchild Semi	SPT7883	10	20	60	9.67	B
Fairchild Semi	SPT7883	50	105	60	9.67	B
Fairchild Semi	SPT7883	70	129	57	9.18	B

[a] A is when ADC is powered off between conversions. B is when current scaling is used to reduce power with sampling rate

noted earlier, this results in a significant increase in power consumption. As a result several techniques have been proposed which allow a pipelined ADC to work without a front-end S/H and thus reduce power.

In previous publications, the front-end S/H was eliminated by relying on the redundancy of the first pipelined stage (e.g. [61–63]). For example, in a 1.5 b/stage architecture the comparator offset in the sub-ADC can be as large as $V_{ref}/4$ (where V_{ref} is the maximum peak voltage of the input). Thus so long as the difference in input operated on by the MDAC and sub-ADC (as shown in Fig. 3.21) is less than $V_{ref}/4$, the effect of sampling skew appears as an input-referred offset on the sub-ADC comparator, and the effect of the offset is eliminated by the redundancy of the first pipeline stage. Hence the front-end S/H can be eliminated without any further modification to the ADC. From equation (5.2–5.21), assuming a sinusoidal input to the ADC with a maximum peak voltage, and assuming no inherent offset in the sub-ADC comparators, the maximum allowable skew time that can be corrected by the redundancy of a 1.5 b stage is $(8\pi f_{in})^{-1}$. Thus high frequency inputs require low skew between sub-ADC and MDAC; for f_{in} = 270 MHz the maximum skew allowable is 140 ps. The practical allowable skew between sub-ADC and MDAC however must also take into consideration the mismatch of the comparators in the sub-ADC (which in a 1.5 b stage are typically made large) as well as a design safety margin – with these factors included the skew must be significantly lower than 140 ps. In a sub-sampled ADC where the input frequencies are necessarily very large, clearly great care [62] and thus increased time (hence increased costs) must be taken in layout and simulation if the first pipeline stage's redundancy is used to eliminate the front-end sample-and-hold such that the system works over all process corner variations and temperatures.

5.4.2 Open-Loop Amplifier Approach

In Section 3.5 it was shown that while feedback around an opamp enables a highly linear and precise gain, it comes at the cost of bandwidth reduction in the closed-loop. Namely by closing the feedback loop, the unity gain frequency of the closed system is that of the open-loop divided by $1/\beta$ (where due to parasitic capacitors β can become small). As a result pipeline stages with large closed-loop gains come at the cost of decreased maximum speed. To overcome this limitation, some researchers have investigated using open-loop amplifiers to reduce power consumption.

In [74] an open-loop technique is used for a 4-bit first pipeline stage in a 12-bit 75 MS/s ADC. Figure 5.11 illustrates the architecture of the 4-bit pipeline stage used in [74].

Rather than using an opamp in closed-loop, an open-loop differential amplifier is used to achieve the desired gain of $\times 8$ in the pipeline stage, where the gain is set approximately by $g_m R_{load} = 8$, where g_m is the transconductance of the differential pair M1–M2, and R_{load} the load impedance seen by the differential pair. As the amplifier in Fig. 5.11 does not require a large DC gain, nor limited by additional parasitics which reduce β in a closed loop system, a power savings of 60% was reported in [74] compared to a closed-loop opamp based approach.

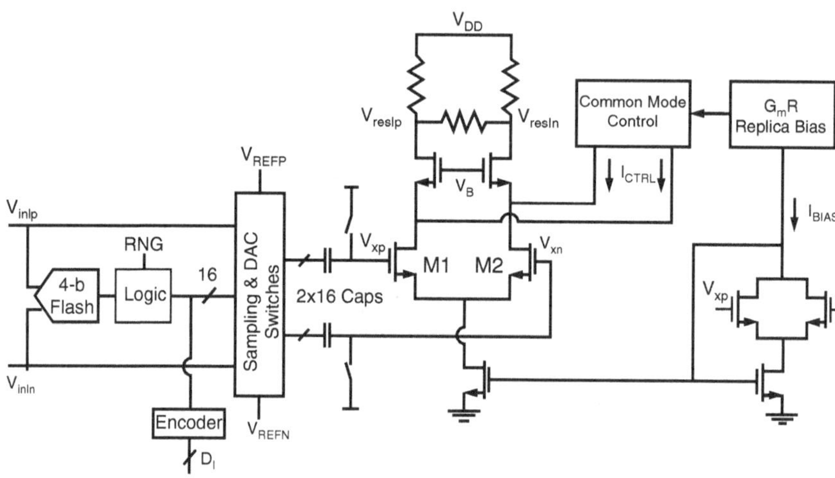

Fig. 5.11 4-Bit pipeline stage using open-loop amplifier (output of stage taken at nodes V_{res1p}, V_{res1n}) (From [74])

A significant advantage of using an open-loop topology is a simple differential pair topology can be used rather than a more time consuming to design opamp. With an open-loop amplifier the designer is free from worrying about stability criteria such as phase margin. Furthermore it is conceivable that with an open-loop amplifier it would be easier to operate with low supply voltages as strictly speaking no stacked devices (which reduce signal swing – e.g. output transistors in a folded cascode) are required to ensure large a DC gain as is the case in opamps.

The tradeoff in using an open-loop topology is that the linearization and process desensitization found in feedback topologies are eliminated. In closed-loop topologies a large open-loop gain from the opamp creates a virtual ground at the input of the opamp and thus significantly reduces the input signal swing to the input differential pair of an opamp, resulting in a very linear closed-loop transfer curve. In an open-loop topology however the input to the differential pair of e.g. Fig. 5.11 has a signal swing one eighth that of the output. As a result the transfer curve of an open-loop approach suffers from a significantly increased non-linearity as shown in Fig. 5.12

Furthermore in a closed-loop system, the gain is typically set by the ratio of capacitors which can be determined to a high precision, whereas in an open-loop approach the gain is set by the product of g_m and R_{load} – a product which varies significantly with process and temperature.

To compensate for the nonlinearity and gain variation in [74], a calibration scheme was developed which corrects the nonlinearity of the open-loop amplifier in the digital domain by using an inverse nonlinearity function, $f^{-1}(x)$, as shown in Fig. 5.13.

In [74] a statistics based digital background calibration scheme was used to estimate the value of $f^{-1}(x)$. The specific details of this calibration scheme are not discussed here, however it is noted that the calibration engine was mathematically intensive due to the large number of calculations involved in digitally estimating the nonlinear function $f^{-1}(x)$ to a precision higher than the 12-bits desired from the ADC. Furthermore, the digital calibration consumed one third the overall power

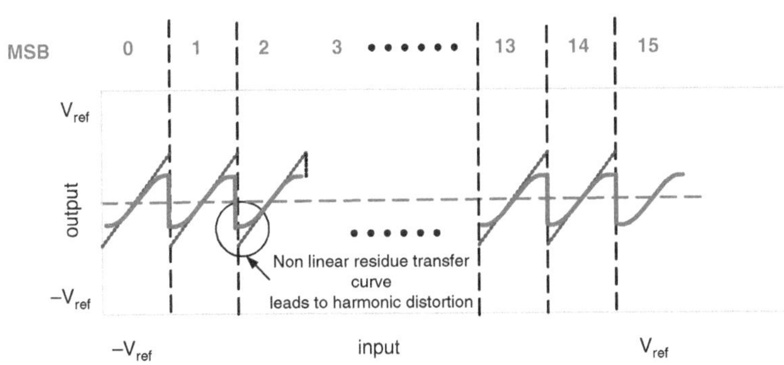

Fig. 5.12 Non-linear transfer curve for residue transfer function – ideal transfer curve shown in *dashed lines*

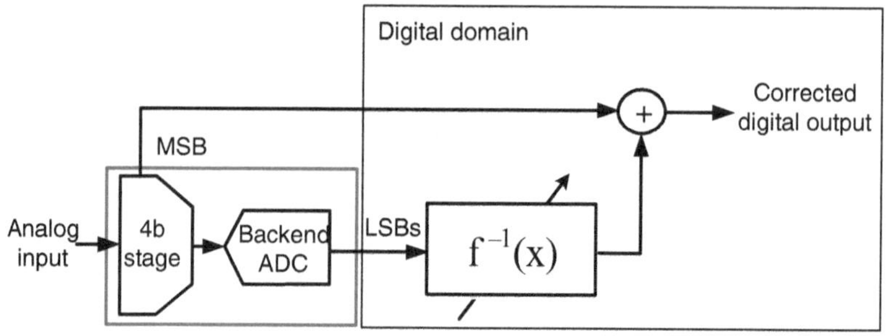

Fig. 5.13 Nonlinearity correction scheme

reported in [74], which was fabricated in a 0.35 μm CMOS process. Thus while the unity gain frequency of the amplifier in an open-loop approach is dramatically increased for a given power, a large amount of power in the digital domain due to the complex DSP is subsequently required to enable the bandwidth enhancement. It is noted however that process scaling favors digital devices from a power and area perspective, hence it could be argued that trading analog power and complexity with digital power and complexity is a favorable one. It is noted that since the calibration scheme in [74] was a statistics based approach, a very large number of clock cycles were required in the calibration (on the order of 10^7 clock cycles to achieve 12-bits of resolution).

5.4.3 Comparator Based Switched Capacitor Circuits

Like open-loop amplifier based pipeline stages, Comparator Based Switched Capacitor (CBSC) circuits [75] overcome limitations from technology scaling by replacing the opamp in closed-loop with a more power efficient structure. In [75], the CBSC technique was used in the MDACs to develop a 2.5 mW, 10-bit 8 MS/s pipelined ADC.

In switched capacitor circuits the fundamental job of the opamp is to provide a virtual ground to ensure accurate charge transfer from sampling to feedback capacitors. Instead of using an opamp to generate a virtual ground, CBSC circuits use a comparator in a feedback loop to effectively emulate the functionality of an opamp with a large DC gain. Figure 5.14 illustrates a CBSC circuit which provides a gain of two.

In Fig. 5.14, during Φ_1, the input is bottom plate sampled on capacitors C_1 and C_2, and node Vx is initialized to a voltage below the value Vx would be if an opamp were used instead of a comparator, i.e. Vx is initialized to be below V cm. During Φ_2, the current source at the output, Io, is turned on and Vout and node voltage Vx subsequently increase in voltage. When node Vx equals V cm, the comparator toggles and turns off the current source Io, and the node voltages at Vx and Vout appear the same as if an opamp were used to create a virtual ground at Vx. Thus the

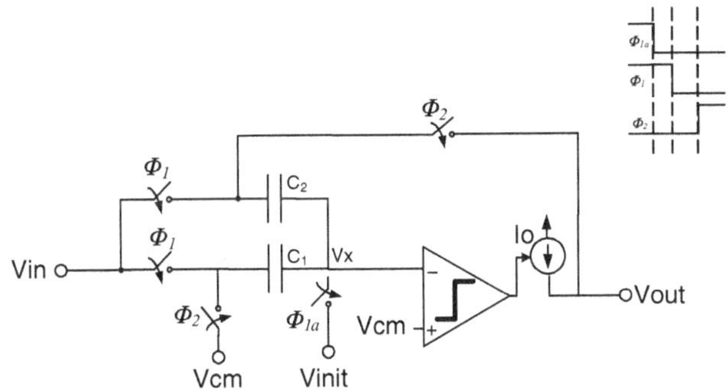

Fig. 5.14 CBSC gain of ×2 circuit

CBSC arrangement implements the same functionality as an opamp based arrangement (i.e. gain of ×2). It is noted that in [75] an additional clock phase is used to ensure a precise charge transfer, however the general principle of the circuit in [75] is the same as shown in Fig. 5.14.

The significant advantage of CBSC is that the topology does not depend on an opamp which would otherwise require a large DC gain and/or large supply voltage to implement a gain of ×2 in a switched capacitor circuit. As comparators can be easily designed even with low supply voltages and with transistors that have low intrinsic gain, CBSC is well suited for implementation in deep submicron technologies. Furthermore in [75] it is shown that the CBSC approach has less inherent thermal noise than an opamp based approach and thus is able to use smaller sampling capacitors, hence have lower power consumption. One of the drawbacks of CBSC is thus far the technique has only been shown in silicon in single-ended form, thus is susceptible to common mode noise which could be very large in an integrated system.

5.5 Summary

This chapter reviewed several state of the art pipelined ADC enhancement techniques. Areas of research most relevant to this book were covered in most detail, namely: rapid calibration, power scalable ADCs, and low power techniques through sample-and-hold removal techniques and non-opamp based pipeline stages.

Part II
Pipelined ADC Enhancement Techniques

Chapter 6
Rapid Calibration of DAC and Gain Errors in a Multi-bit Pipeline Stage

6.1 Overview

In Section 5.2.6 rapid calibration techniques were shown to be of great interest in industry. Furthermore it was discussed that thus far rapid calibration techniques only correct gain error in 1.5 b/stage based topologies. In this chapter a technique to digitally rapidly correct for both DAC and gain errors in the multi-bit first stage of an 11-bit pipelined ADC is presented. The architecture [2, 3] uses a split-ADC based approach. A proof-of-concept prototype of the digital background scheme was fabricated in a 1.8 V, 0.18 μm CMOS process, where measured results show the calibration scheme is able to improve the peak INL of an 11-bit 45 MS/s ADC from 6.4 LSB to 1.1 LSB after calibration. Measured results also show the SNDR/SFDR of the ADC improves from 46.9/48.9 dB to 60.1/70 dB after calibration. It will be seen that the calibration is achieved in approximately 10^4 clock cycles, which is more than $\times 100$ faster than previously published techniques which correct for both gain and DAC errors.

The organization of key sections in this chapter is as follows: Section 6.2 discuses the motivations of the work. Section 6.3 discusses the architecture used to implement rapid calibration of both DAC and gain errors in a multi-bit stage of a pipelined ADC. Section 6.4 discusses the circuits used to implement the ADC of this work. Section 6.6 discusses measured results of a prototype fabricated in a 1.8 V, 0.18 μm CMOS process. Section 6.7 concludes the chapter.

6.2 Motivation

In Sections 3.4 and 3.7 it was detailed how ADC accuracy was limited by imperfections such as low opamp DC gain and capacitor mismatch. In Section 5.2 various techniques to measure and correct gain and capacitor mismatch were described. In particular the split-ADC topology was discussed as a useful approach to significantly reduce calibration time. The published reports of the split-ADC approach

I. Ahmed, *Pipelined ADC Design and Enhancement Techniques*,
Analog Circuits and Signal Processing,
DOI 10.1007/978-90-481-8652-5_6,

thus far however have only showed successful implementation in 1.5 b/stage based topologies. In this work the split-ADC technique is expanded to also rapidly correct DAC errors in a multi-bit pipelined ADC stage. As discussed in Section 5.2.3, since gain errors appear as a constant DAC error, the DAC calibration scheme discussed in this chapter also corrects gain errors in a pipelined stage. Hence the topology of this chapter is a more general or superset solution to those discussed previously in literature.

6.2.1 Why Are DAC Errors Important to Correct?

As discussed in Section 3.6 ADCs with a low desired thermal noise floor requires large capacitors to suppress thermal noise. Since capacitor mismatch decreases with increasing capacitor area, it could be argued that in high resolution ADCs the DAC errors are automatically addressed by virtue of using large sampling capacitors. In many applications however, the ADC does not need a very low thermal noise floor, however does need a very linear transfer characteristic – as discussed in Section 2.2, OFDM applications are a prime example of this. From Section 2.2 it was shown that for each individual signal in an OFDM spectrum, the SNDR was limited primarily by in-band harmonic distortion, rather than the thermal noise floor, hence small sampling capacitors in theory could be used to minimize power consumption. On the other hand, as shown in Section 3.7 large sampling capacitors are still required in OFDM type applications to suppress the effects of capacitor mismatch. From Section 5.2.3 it was shown that calibration techniques could be used to compensate for the effect of capacitor mismatch and thus enable a designer to use small sampling capacitors for an ADC in an OFDM type application. However it was shown in Chapter 5 that state of the art DAC calibration techniques all use statistical based schemes and thus suffer from lengthy calibration times, which from a production standpoint was shown in Section 5.2.6 to be unfavorable. Thus this the goal of this work was to develop a DAC calibration scheme, hence enabling low power ADCs in certain applications, where the calibration could be achieved in a very short time interval, and thus be very useful in industry.

6.3 Rapid DAC + Gain Calibration Architecture

The calibration scheme of this work uses a dual-ADC approach (ADC A and ADC B), to simultaneously process in parallel the same analog input as shown in Fig. 6.1. The final ADC output is generated by the average of the two ADC outputs, thus each ADC is designed with half the total capacitance, hence half the power and area of the overall ADC to meet thermal noise requirements [49].

From Fig. 6.1 ADC A and B are identical except in each ADC the residue transfer function in the first stage is horizontally offset from the other by approximately half a MSB.

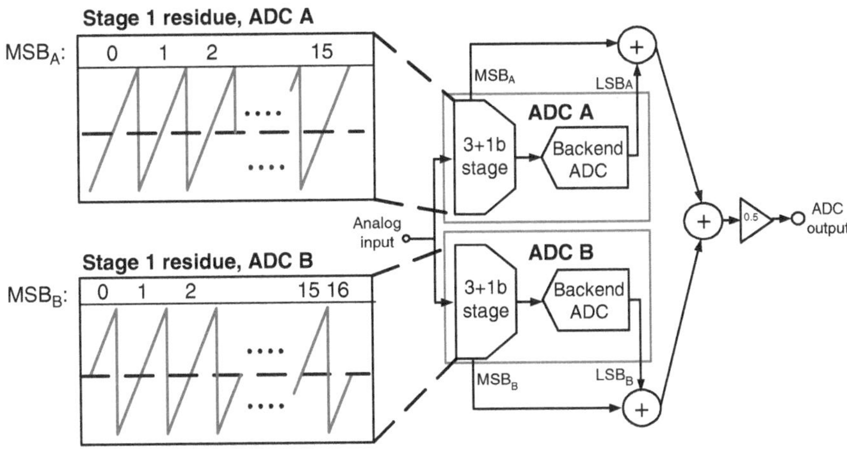

Fig. 6.1 Dual-ADC approach of this work

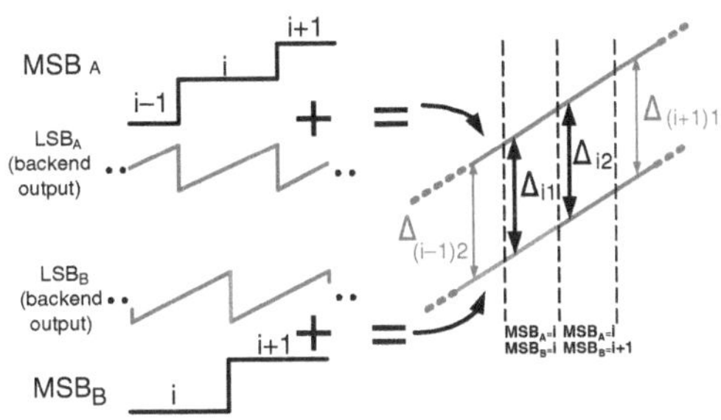

Fig. 6.2 Transfer curves of first stage (MSB), backend ADC (LSB) and total ADC outputs from each split-ADC with no errors

6.3.1 *Measurement of Missing Codes Due to DAC and Gain Errors*

Figure 6.2 illustrates the transfer function of key outputs from each split-ADC. From Fig. 6.2 if the analog input to the ADC is such that $MSB_A = i$, then MSB_B is either i or $i + 1$. The offset between the digital outputs of ADCs A and B for the range of analog inputs where $MSB_A = i$ and $MSB_B = i$ is denoted Δ_{i1}, and Δ_{i2} where $MSB_A = i$ and $MSB_B = i + 1$ respectively, as shown in Fig. 6.2. In an ideal ADC without DAC or gain errors the difference between split-ADC outputs is constant regardless of the analog input, thus $\Delta_{i1} = \Delta_{i2}$. With ideal ADCs A and B $\Delta_{i1} = \Delta_{i2} = 0$, however the offset is shown as a non-zero constant in Fig. 6.2 for clarity of illustration.

If however DAC and gain errors are included, as shown in Fig. 6.3, each split-ADC incurs unique missing codes wherever an MSB changes.

As the MSB transitions between each split-ADC are staggered however, ADC A does not incur an error in the first pipeline stage for the same range of analog inputs as ADC B. Thus the digital output of ADC A can be used as an *ideal reference* to measure the errors of ADC B. The difference between Δ_{i1} and Δ_{i2} precisely gives

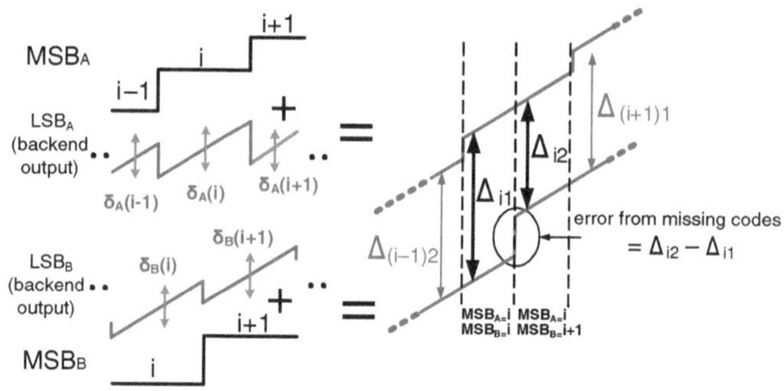

Fig. 6.3 Transfer curves of key ADC outputs with gain, DAC errors included

the error due to missing codes that occurs when MSB_B changes from i to i + 1 as shown in Fig. 6.3. To minimize the effect of zero mean error sources such as thermal noise, Δ_{i1} and Δ_{i2} are averaged before subtraction using a simple first order IIR filter with transfer function $\mu/[1-(1-\mu)z^{-1}]$. In this work $\mu = 1/64$ was used, where by implementing μ as a power of 2, multiplication by μ can be implemented using simple bit shifts. In a similar manner the unique error due to missing codes at all other MSB transitions can be measured for ADC B.

Errors due to missing codes for ADC A are measured by noting that $\Delta_{i2}-\Delta_{(i+1)1}$ is the error due to missing codes in ADC A when MSB_A changes from i to i + 1 as shown in Fig. 6.3. Hence the error due to missing codes in ADC A can be determined using already measured values $\overline{\Delta_{i2}}$ and $\overline{\Delta_{(i+1)1}}$. Errors due to missing codes at all other MSB transitions in ADC A are measured using an identical extension as done for ADC B.

6.3.2 Correction of Missing Codes

With the errors from missing codes at each MSB transition measured, each ADC is corrected by shifting each ADC's digital output as a function of the MSB such that overall transfer function of each ADC is free from missing codes due to errors in the first stage as shown in Fig. 6.4 (same done for ADC A).

Rapid calibration is achieved as $\overline{\Delta_{iB}}$ is highly correlated with the number of missing codes; only a small number of clock cycles are required to average out the effects of zero-mean noise. As long as the input is sufficiently busy to generate a sufficient number of estimates of Δ_{i1}, Δ_{i2}, for all i, there is no constraint on the shape of the input signal to the ADC. In contrast statistical techniques use statistical correlations which require many output samples to extract similar information. The full topology of the ADC including digital calibration is shown in Fig. 6.5.

It is noted that the approach of this work is similar to background calibration techniques where a more accurate but slower ADC is used in parallel with the ADC under calibration (e.g. [55]), where by adaptively equalizing the outputs of the fast ADC to the output of the slow ADC calibration can be achieved. In this work since

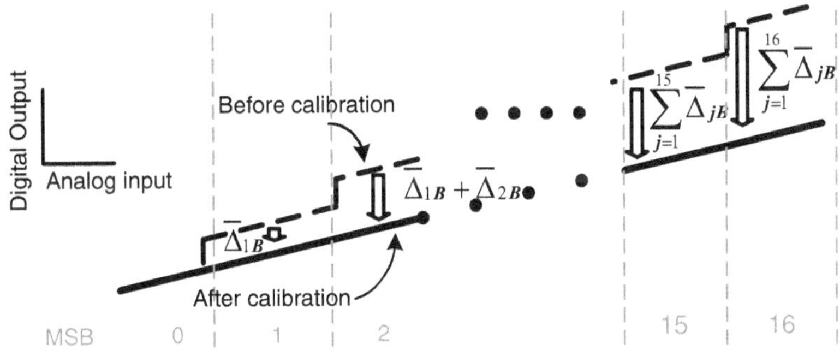

Fig. 6.4 Illustration of how correction terms for ADC B are derived from estimates of missing codes (correction topology of ADC A is similar)

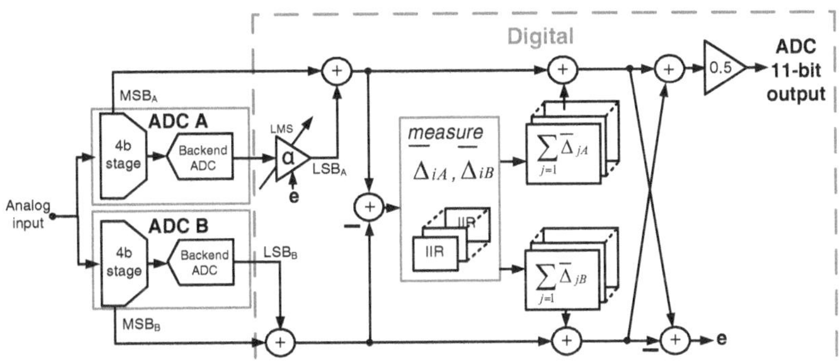

Fig. 6.5 Full ADC topology of this work

the residue transfer function of one of the split-ADCs is offset from the other, ADC A does not suffer an error in the first stage for the same range of inputs as ADC B, thus one ADC can be used as an ideal reference for the other eliminating the need for one of the ADCs to be more accurate than the other. Hence there is no need to trade higher accuracy with lower sampling rates in the second ADC; both ADCs can operate at the same speed and both ADCs can be used to digitize the analog input. Thus the power of the additional ADC also goes towards lowering the noise floor in the digital output, unlike [55] where the additional ADC (since it operates slower) only aids the correction scheme. Furthermore using the technique outlined in this work, both ADCs are calibrated whereas in [55] only one ADC is.

6.3.3 Mismatch Between ADCs

Due to random mismatches between the split-ADCs, the first stage of ADC A in Fig. 6.1 will have MSB transitions that are not exactly half MSB apart, as well as a different input referred offset and different stage gain than ADC B.

The calibration scheme of this work only requires the MSB transitions between ADC A and B to be non-overlapping, thus a precise definition of the offset between ADCs is not necessary. As long as the comparators which define the MSB transitions for each split-ADC are designed to have an offset within half MSB no errors are produced by the calibration scheme.

Constant offset between the outputs of ADC A and B appears as a common-mode shift in both $\overline{\Delta_{i1}}$, and $\overline{\Delta_{i2}}$. Since the number of missing codes at each MSB transition is measured by subtracting $\overline{\Delta_{i2}}$ from $\overline{\Delta_{i1}}$, the common mode is eliminated and thus input-referred offsets of each split-ADC have no impact in the calibration scheme (under the practical assumption that the offsets are not large enough to saturate the output of the pipeline stages).

To account for an overall gain mismatch between the two ADCs an LMS adaptive gain term (α in Fig. 6.5) is also included [49] which scales the backend code of ADC A so as to keep the outputs of ADC A and B parallel. As ADC B provides an ideal reference for ADC A, the error signal used for the LMS adaptation (which is formed by the difference of the two ADC outputs) is highly correlated with the gain error between ADCs, thus steady state convergence of α occurs within a relatively short time interval. To ensure stability, the adaptation of α is designed to converge more slowly than the loop which measures and corrects the gain and DAC errors of the first pipeline stage.

Ultimately, the accuracy of estimation in errors due to missing codes in the first stage is limited by missing codes and distortion in the backend, which although not addressed in this work can be minimized by also calibrating the backend stages.

6.3.4 Simulation Results

To evaluate the functionality of the calibration technique discussed in this chapter, a model of the system was implemented in Matlab Simulink. Gain and DAC errors were modeled in the first 3 + 1-bit pipeline stage, along with thermal noise, as shown in Fig. 6.6. The thermal noise floor was designed such that the input referred noise floor was approximately at the 11-bit level.

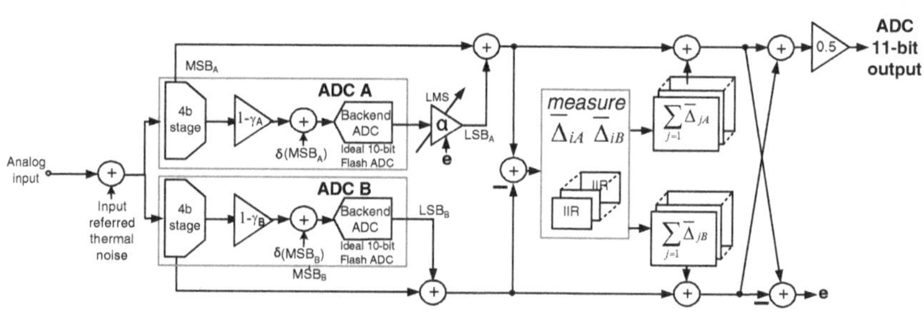

Fig. 6.6 Matlab Simulink test setup for simulation verification

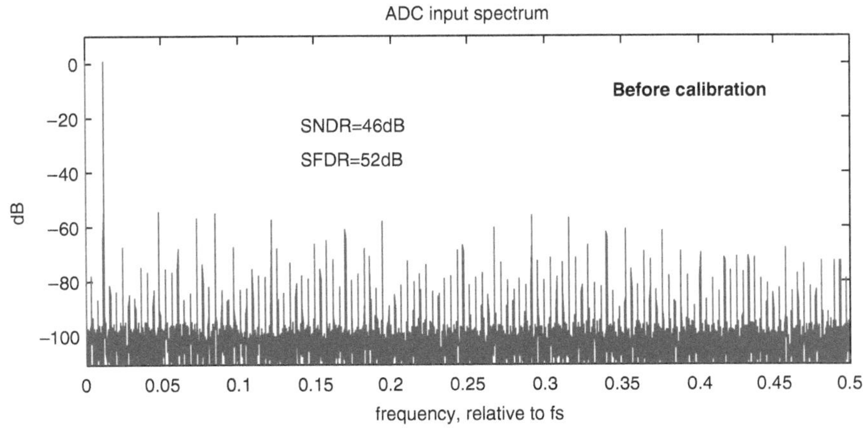

Fig. 6.7 Simulation results with 1% gain and DAC error – before calibration

Each dual-ADC in Fig. 6.6 was initialized with a different static gain error γ, and unique DAC errors Δ at startup, such that each ADC path was not identical. To concentrate on the effectiveness of the calibration scheme, the backend ADC was modeled as an ideal 10-bit Flash ADC. The digital calibration engine implemented in Simulink emulated the topology shown in Fig. 6.5, where care was taken to ensure all digital correction terms had no more than 14-bits.

Figure 6.7 shows the FFT plot of the ADC (before calibration) when the first stage in each pipeline stage was designed to have a relative gain error of $\sim$1% and a relative capacitor mismatch ratio also of $\sim$1%. The input to the ADC was a sinusoid at 1/200th the sampling rate so that all key harmonics could be shown without being aliased.

Figure 6.8 illustrates the FFT after calibration where the SNDR/SFDR is seen to dramatically improve to 66/85 dB.

Figure 6.9 shows the evolution of ADC SNDR and SFDR with the number of clock cycles, where it is seen that a 23 dB in SFDR and 20 dB in SNDR improvement can be achieved in less than 1.5×10^4 clock cycles.

To verify that the system worked with random inputs, an additional simulation was performed where a uniform random input was applied to the ADC for 1.5×10^4 clock cycles and the error terms measured. After 1.5×10^4 clock cycles the error measurement was frozen and a sinusoid applied to the input of the ADC where an FFT was subsequently taken of the sinusoidal digital output. Figure 6.10 demonstrates the FFT of the ADC after calibration where it is verified that calibration is achieved even with a random input.

Figure 6.11 shows the evolution of ADC SNDR/SFDR when calibration correction terms are derived with a random input.

Similar experiments were conducted with different input sources (e.g. triangle wave, saturating input, etc.) and in all cases calibration was achieved so long as the input was sufficiently busy to excite each correction term.

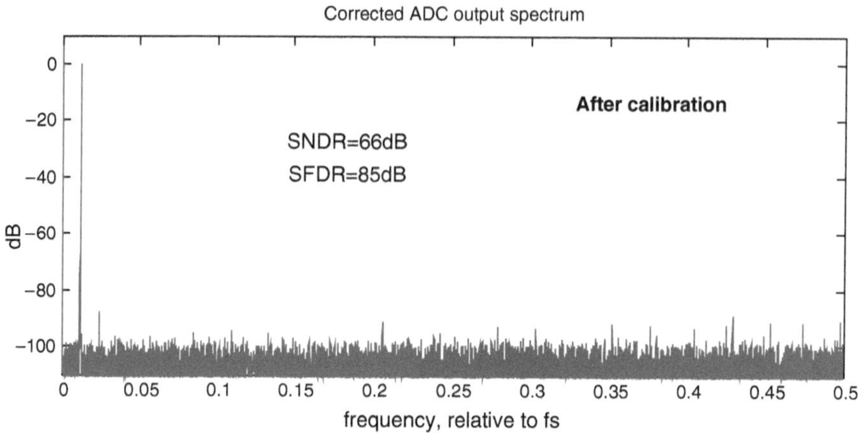

Fig. 6.8 Simulation results with 1% gain and DAC error – after calibration

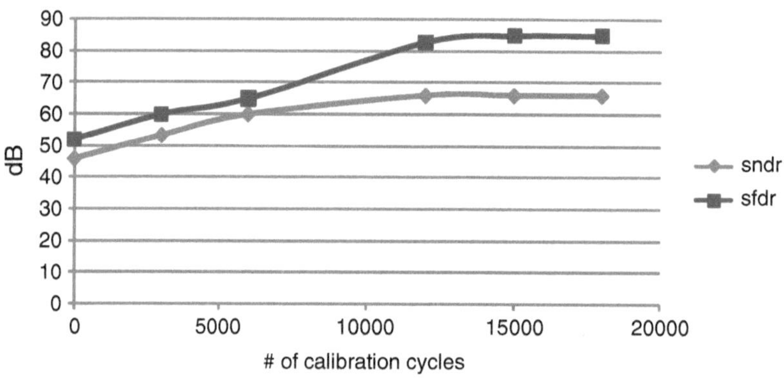

Fig. 6.9 SNDR/SFDR improvement with calibration cycles in Simulink model

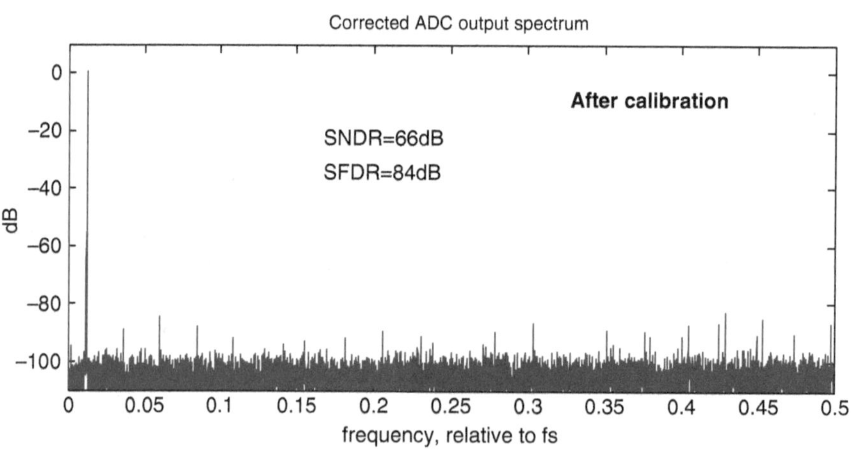

Fig. 6.10 FFT of ADC after calibration driven by uniform random input for 1.5×10^4 clock cycles

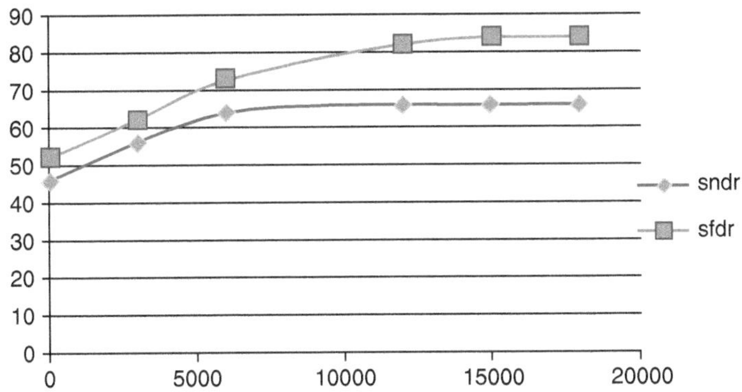

Fig. 6.11 SNDR/SFDR improvement with calibration cycles when ADC driven by a uniform random input

From Figs. 6.7–6.11 it is clear that even with a very large capacitor and gain mismatch on the order of 1%, the calibration scheme is able to provide both effective and rapid correction of circuit non-idealities, thus justifying a fabricated prototype.

It should be noted that the adaptation coefficients were determined heuristically. The value of μ ultimately selected in this work of 1/64 was found to adequately achieve a sufficient amount of error correction for the design specifications of this work for a variety of different inputs and reasonable mismatch levels. A detailed analysis of the impact of correction terms in the architecture of this chapter is deferred as future work.

6.4 Circuit Implementation

To evaluate the architecture of this chapter a 45 MS/s 11-bit linear pipelined ADC was constructed in a 1.8 V, 0.18 μm CMOS process. From Section 6.3.4, although the calibration technique of this chapter could enable much higher resolution ADCs, 11-bit resolution was ultimately chosen as to achieve a very high linearity for a mid-high speed ADC (i.e. >40 MS/s), a lengthy design of a PCB, and IC package would have to be undertaken to ensure sufficient suppression of noise sources (e.g. digital I/O noise, LC bond wire oscillations) in the proof-of-concept prototype. Since only the first stage errors are corrected using the calibration technique presented in this chapter, a 3 + 1 bit first stage was used so that the non-idealities of the backend ADC would be small and not require calibration. The thermal noise floor of the ADC was designed to be near the 11-bit level for a 1.3 V p-p input signal. Although a specific application was not targeted, a quick survey of industry data sheets of ADCs with 11-bit linearity with sampling rates around 45 MS/s show for example: CCD imaging, video, portable instrumentation, IF/baseband communications, cable modems, wireless LAN, and medical ultrasound as potential applications.

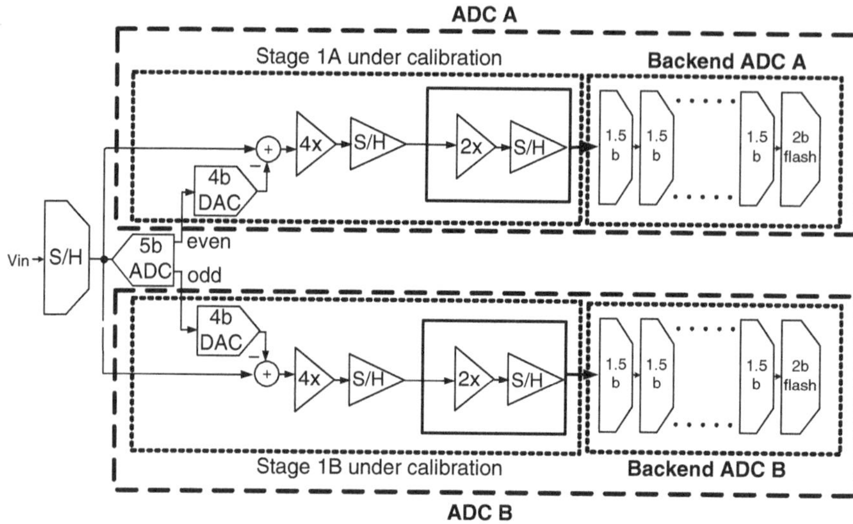

Fig. 6.12 Analog portion of ADC topology in detail

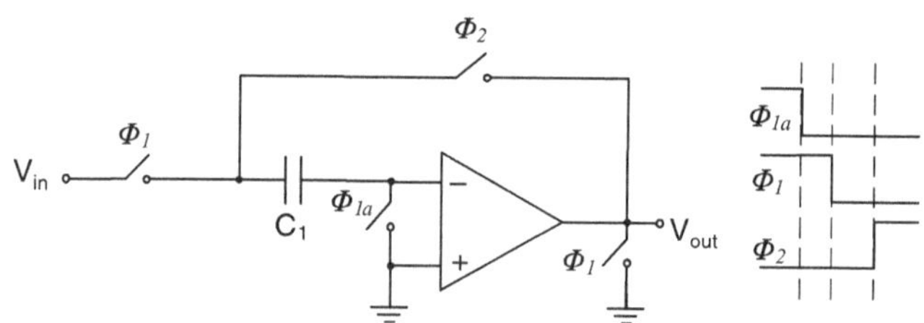

Fig. 6.13 Sample-and-hold topology (implemented fully-differentially in this work)

Figure 6.12 shows the architecture of the analog portion of the 11-bit pipeline ADC of this work. Scaling was used in the first three pipeline stages to reduce power consumption [16]. An additional three bits were added to the backend in each split-ADC to improve the accuracy in error estimation. As the last stages in a pipelined ADC consume only a small fraction of the total power, adding extra stages to reduce quantization noise has a minimal impact on ADC power.

6.4.1 Front-End Sample-and-Hold

To ensure both ADCs operate on the same analog input, a conventional flip-around front-end sample-and-hold [25] was used and is shown in Fig. 6.13. The flip-around topology has the advantage of a feedback factor near unity hence is power efficient. The sample-and-hold was designed such that the entire ADC had approximately an

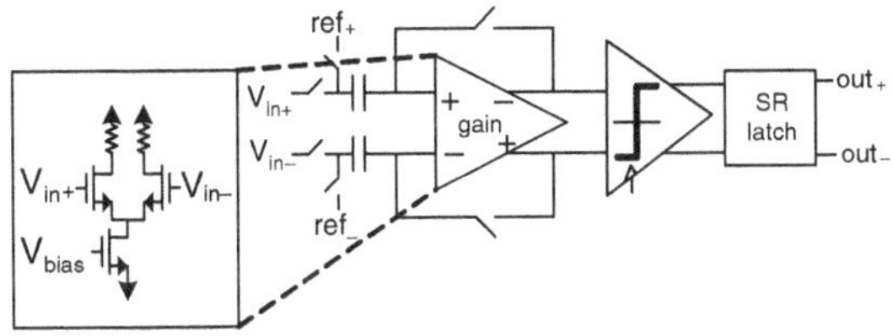

Fig. 6.14 Comparator topology used in 5-bit flash sub-ADC

11-bit input referred thermal noise floor relative to a 1.3 V peak-to-peak differential input sinusoid.

6.4.2 5-Bit Flash ADC

The MSB transitions for each split-ADC were generated by a single 5-bit flash ADC where the even numbered outputs were used for ADC A and odd numbered outputs for ADC B. The 5-bit flash ADC was implemented using an array of comparators where a single comparator is shown in Fig. 6.14.

Each comparator used a preamp before the slicer to reduce kickback noise from the slicer on to the reference voltages which were implemented with a resistor string. A switched capacitor based topology [10] was used to implement the threshold of the comparator as well as reduce offset from the preamplifier.

From Fig. 6.12 it is noted that one of the overheads of a split-ADC topology is an increase in resolution of the sub-ADC of the stage under calibration by 1-bit (to ensure the two ADCs do not have overlapping MSB transitions). However since the power of the sub-ADC is only a small fraction of overall ADC power, increasing the resolution of the sub-ADC by 1-bit results in only a small overall increase in analog power. Furthermore for low resolutions in the sub-ADC, as is the case in this work, the power of the sub-ADC is not limited by thermal noise – thus increased resolution can be obtained with only a marginal increase in power so as to sufficiently minimize comparator offsets.

6.4.3 4-Bit MDAC

The topology of the multi-bit MDAC in the first stage of ADC A is shown in Fig. 6.15. The MDAC for ADC B is identical except slightly modified to account for a half MSB horizontal offset.

As capacitor mismatch is a process variation, its impact can only be properly observed by testing thousands of chips for random variation. To minimize the number of ICs tested each sampling capacitor in the first stage of each split-ADC was made individually digitally programmable with 0.5%, 1% and 5% deviations

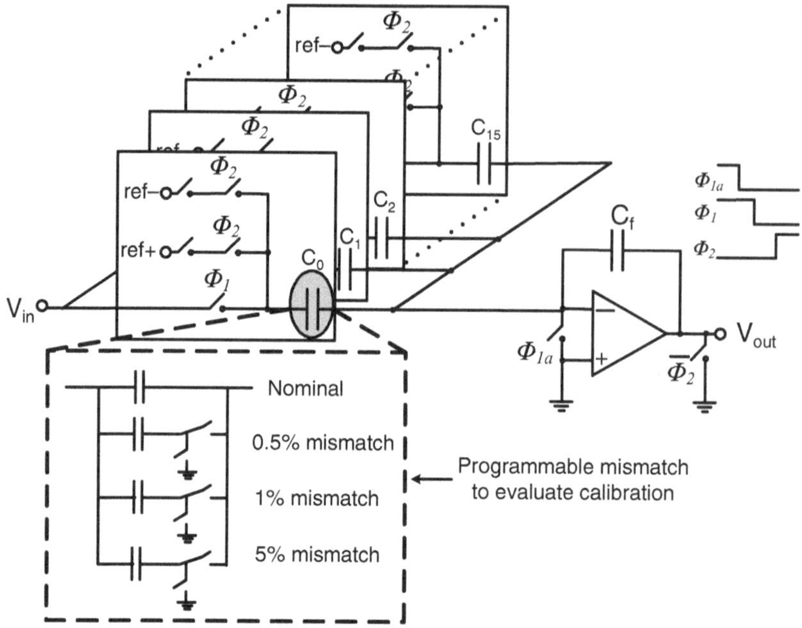

Fig. 6.15 Topology of first stage MDAC (implemented fully-differentially in this work)

from nominal values as shown in Fig. 6.15. Thus the effectiveness of the calibration scheme of this work could be very quickly evaluated by programming different mismatch profiles in the first stage of each split-ADC. The control bits to set the mismatch profile of the first stage in each ADC were set via an on-chip shift-register.

Since the calibration scheme of this work also corrected gain errors, the DC gain of the opamp in Fig. 6.15 was designed with the minimum DC gain required to adequately suppress distortion from the opamp. The opamp, which had a DC gain of ~50 dB, was implemented using a one-stage p-input folded cascode topology, and is illustrated in Fig. 6.16. Common mode feedback was implemented using a switched capacitor based approach [10].

In a 4-bit pipeline stage including 1-bit of redundancy, the ideal closed-loop gain of the MDAC is ×8. It is noted however that the feedback factor β of the MDAC of Fig. 6.15 is given by

$$\beta = \frac{C_f}{\sum_{i=0}^{15} C_i + C_p + C_f}, \tag{6.1}$$

where C_p is the parasitic capacitance at the input of the opamp. The time constant of the settling time τ of the MDAC given by:

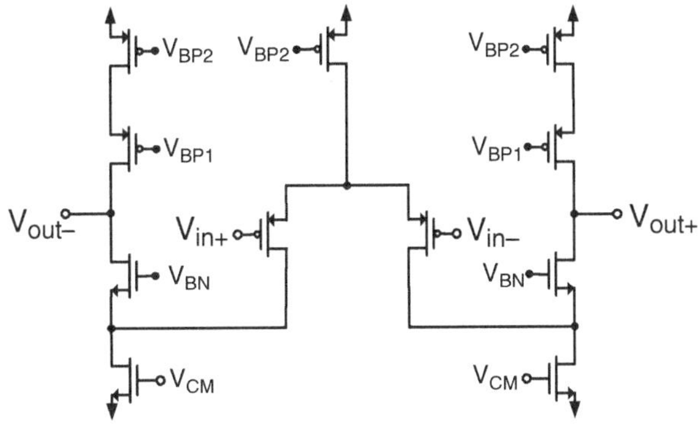

Fig. 6.16 Opamp used in first stage MDAC

$$\tau = \frac{1}{\beta \omega_{ta}} \tag{6.2}$$

where ω_{ta} is the open-loop unity gain frequency of the opamp. Thus as the closed-loop gain becomes larger C_f becomes smaller and the feedback factor β becomes more sensitive to variations in parasitic capacitance at the input of the opamp. Furthermore, small feedback factors require the opamp to have a large open-loop unity gain frequency to achieve fast closed-loop settling as noted in Section 3.5. Opamps requiring high unity gain frequencies can approach the limits of their technology, making the design less power efficient. In this design although measurements are presented at 45 MS/s, the design was targeted for a $\sim\times 2$ higher sampling rate. Limitations in the test setup restricted measurements to half the target sampling rate. Thus in this design to reduce the open-loop opamp unity gain frequency (with the higher sampling rate in mind) as well as sensitivity to parasitic capacitors, the closed-loop gain of the MDAC was designed to be $\times 4$ rather than $\times 8$. To ensure the backend ADC had a full scale analog input, a pipelined $\times 2$ gain stage was inserted between the 4-bit MDAC and backend ADC as shown in Fig. 6.12. The gain error of the $\times 2$ stage is effectively lumped into the gain error of the 4-bit MDAC, hence its non-idealities are also calibrated.

It is noted that for ease of implementation in this prototype the calibration scheme requires the input to be sufficiently busy to excite each sampling capacitor to achieve full calibration. However it is conceivable to relax the input signal swing requirement by rotating the sampling capacitors in a known sequence not correlated with the analog input, so that (e.g.) different sampling capacitors can be used to sample a DC input, hence exercise the range of different MSB transitions.

6.4.4 Backend Pipelined ADC

The backend ADC was implemented using a cascade of nine standard 1.5 b/stage pipeline stages followed by a 2-bit flash ADC as shown in Fig. 6.12. Gain-boosted

[21] folded cascode opamps with a DC gain of 95 dB were used in stages 2 and 3 to minimize missing codes in the backend. The remaining pipeline stages used folded cascode based opamps with DC gains of 50 dB.

6.4.5 Digital Calibration

To enhance flexibility in the test setup, the digital calibration engine was implemented off-chip, where the digital outputs of each split-ADC stage were taken off-chip and imported into Matlab via a logic analyzer. The digital outputs of the fabricated chip were input to a Simulink model of the digital calibration scheme outlined in this work. Since there is no feedback or direct interaction with the analog portions of the ADCs in the calibration scheme of this work, operating on the off-chip digital outputs verifies the calibration scheme without any loss of generality.

Since the missing codes at each MSB transition for both split-ADCs are required to be measured, 16 + 15 = 31 unique missing codes must be measured and estimated. Thus in a potential on-chip digital implementation each Δ_i is required to be passed through an IIR filter, and the digital output offset by each missing code as shown in Figs. 6.17 and 6.18.

It is noted that each averaging block in Fig. 6.18 is only updated once every time its particular MSB bit is excited, i.e. has low activity. Thus the power of each path in Fig. 6.18 is fairly low since with a reasonably random input each MSB bit is excited at a small fraction of the overall sampling rate.

6.5 Testing

Sections 6.5.1 and 6.5.2 discuss the test setup and measured results of the prototype fabricated in a 1.8 V, 0.18 μm CMOS process.

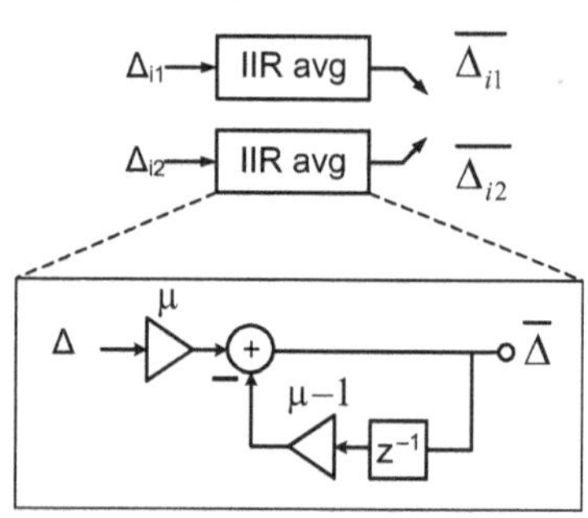

Fig. 6.17 Detail of IIR filter blocks used in 'estimate error' block of Fig. 6.18

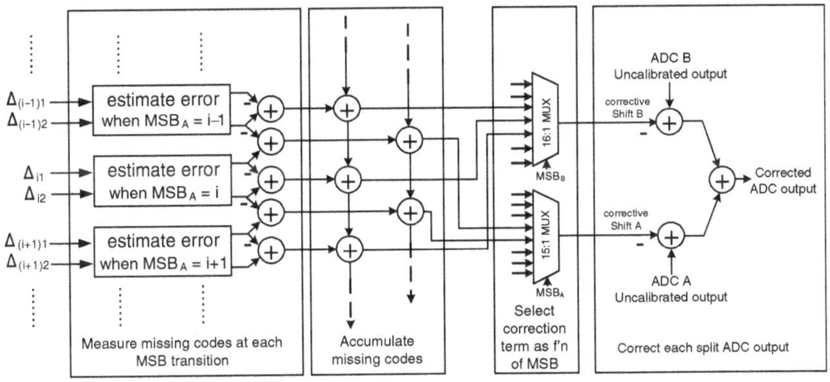

Fig. 6.18 Digital implementation of calibration

6.5.1 PCB

A four layer FR4 dielectric PCB board with a minimum 6 mil trace was designed and constructed for the device under test as shown in Fig. 6.19. Separate Power planes were used to isolate the analog, digital, I/O, and board power supplies. A differential input was generated using a 1:1 turns ratio Minicircuits transformer matched to 50 Ω. Reference voltages were generated by passing the output of a resistive voltage divider through an opamp (LM7301) in a unity gain buffer configuration. To maintain constant supply voltages, all voltage supplies for each power plane were generated through regulators (LM337, LM1117), and heavily decoupled with capacitors. As the ADC utilized a constant current biasing scheme, an off-chip adjustable resistor was used as the master current source.

6.5.2 Test Setup

A test setup as shown in Fig. 6.20 was used. Sinusoidal inputs were generated using an HP 8664A function generator. A Minicircuits low pass filter was used to minimize harmonic distortion from the function generator such that the sinusoidal input to the ADC had an SNDR of well over 11-bits. A 45 MHz clock was derived from a HP8130A pulse/pattern generator. The serial shift register was loaded via a parallel port connection to a PC, where a Matlab script was executed to load the appropriate bits. The output bits of each pipeline stage were captured using a Tektronix TLA714 logic analyzer, capable of capturing 65,536 points at a time. An Agilent E3620A Dual output DC power supply was used to provide positive and negative voltages to the voltage regulators on the PCB. As discussed in Section 6.4.5 the digital calibration was implemented off-chip in a behavioral Matlab model.

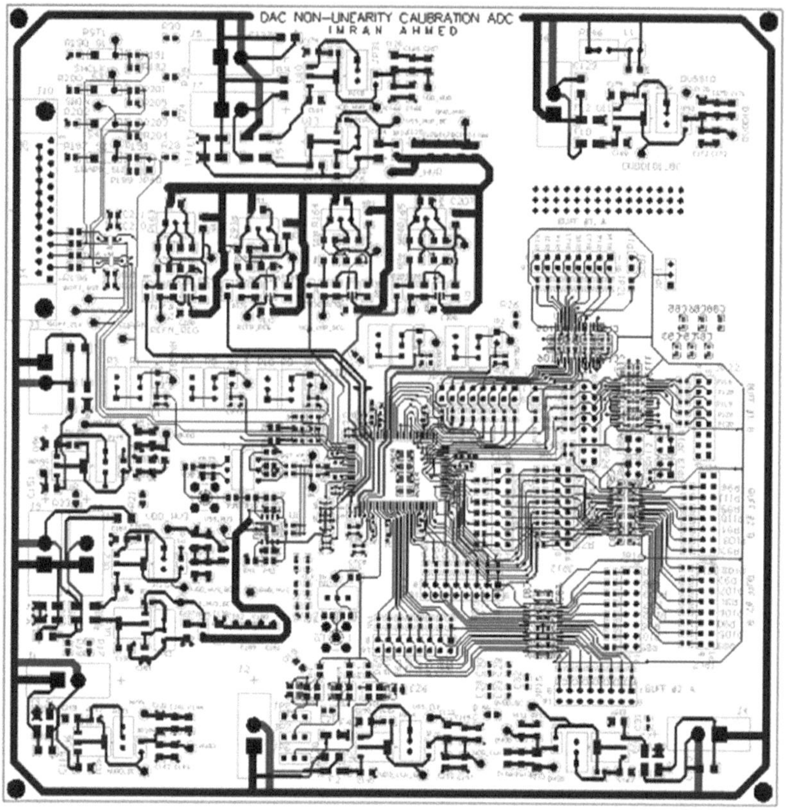

Fig. 6.19 Custom PCB layout

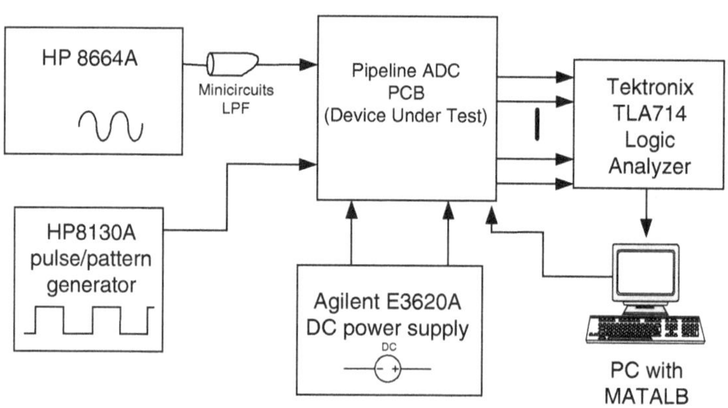

Fig. 6.20 Test setup for rapid DAC calibration ADC

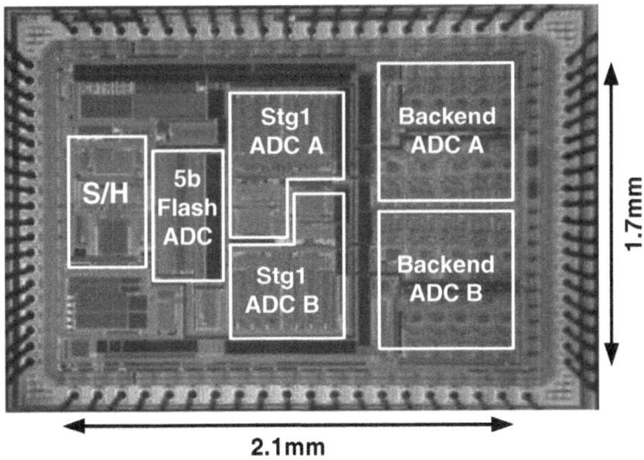

Fig. 6.21 Micrograph of fabricated IC in 1.8 V, 0.18 μm CMOS

6.6 Measured Results

A prototype of the 45MS/s ADC architecture as shown in Fig. 6.21 was fabricated in a 1.8 V, 0.18 μm CMOS process. The area of the analog core was 2.1 × 1.7 mm = 3.57 mm^2. It is noted that the area of the first pipeline stage in each ADC had a large overhead due to the programmability of each sampling capacitor. In a practical design which does not have the additional test circuitry, the area of the first pipeline stage could be significantly reduced.

6.6.1 INL/DNL Plots

Figures 6.22 and 6.23 show the INL and DNL of the ADC before and after calibration, where each DAC element in each split-ADC was programmed with a mismatch of either: 0%, 0.5%, 1% or 5%.

Using the calibration scheme of this work, the INL was improved from +6.1/−6.4 LSB to +1.1/−1 LSB after calibration, and DNL from +1.1/−0.4 LSB to +0.45/−0.4 LSB. It is noted that the residual INL errors after calibration are due primarily to distortion from the backend ADCs, as well as distortion from the front-end sample-and-hold which sets the best achievable linearity for both ADCs.

6.6.2 SNDR/SFDR Plots

Figure 6.24 shows an FFT of the ADC output before and after calibration with a 1.3 V p-p input sinusoid at 2.39 MHz. The SNDR/SFDR of the ADC was improved from 46.9/48.9 dB to 60.1/70 dB.

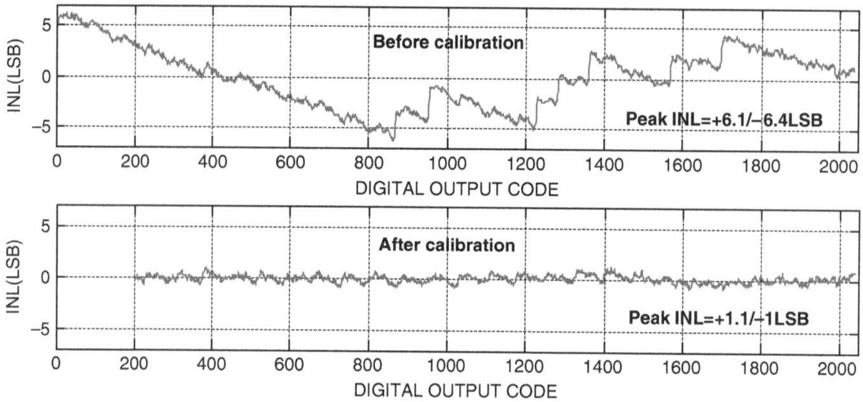

Fig. 6.22 INL before and after calibration, $f_s = 45$ MS/s (LSB @ 11-bit level)

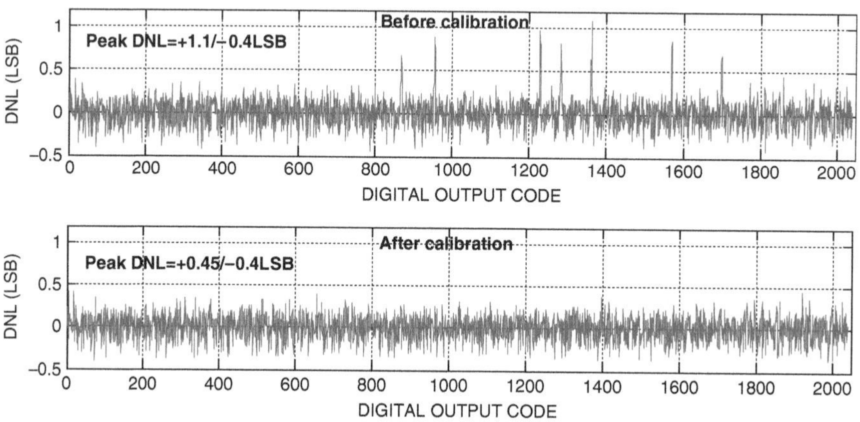

Fig. 6.23 DNL before and after calibration (LSB @ 11-bit level)

The calibration technique was verified with different programmed mismatches in the first stage as well as different full scale inputs (e.g. sinusoidal, random). In all cases successful background calibration was attained in a short time interval. Figure 6.25 illustrates the variation of SNDR and SFDR with input frequency before and after calibration, where the fall-off of accuracy for higher input frequencies was attributed to the input switch in the front-end sample-and-hold.

6.6.3 Calibration Time

The improvement of ADC SNDR and SFDR is plotted versus total number of calibration cycles in Fig. 6.26, where it is shown that steady state is attained within

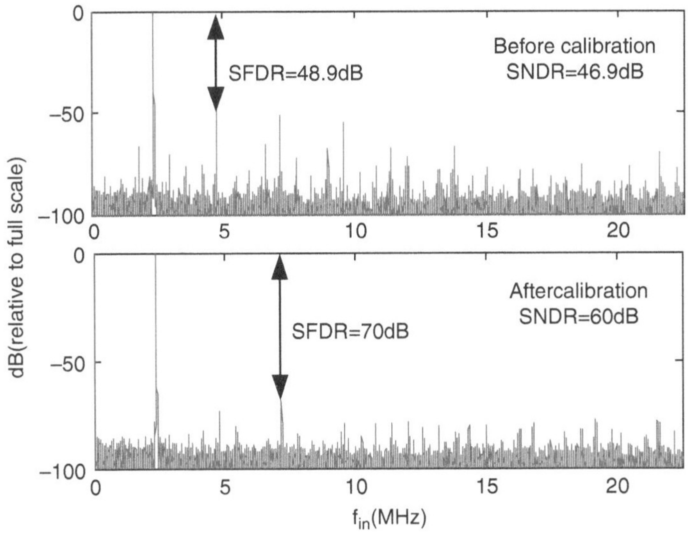

Fig. 6.24 FFT of ADC output before and after calibration

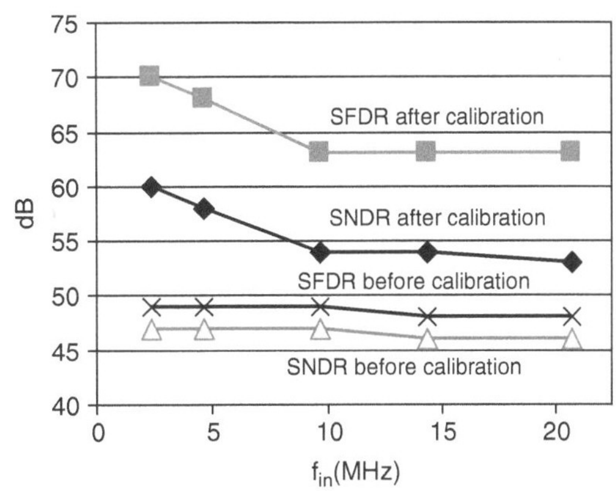

Fig. 6.25 Variation of ADC SNDR, SFDR with input frequency, before and after calibration

$\sim 1 \times 10^4$ clock cycles or effectively 0.22 ms (with $\mu = 1/64$). From [49] it is noted that if a statistics-based background calibration technique were used $\sim 4 \times 10^6$ clock cycles would be required to correct only the gain error. In this work both gain and DAC errors are corrected in less than 1/100th the total number calibration cycles of prior statistics based approaches. The plot of Fig. 6.26 was obtained by running the calibration for a given time and then freezing the calibration coefficients. Once the coefficients were frozen an FFT of the ADC output was taken with a sinusoid applied to the ADC input, and the SNDR/SFDR measured. This process was repeated for a range of total calibration cycles, yielding the plot shown in Fig. 6.26.

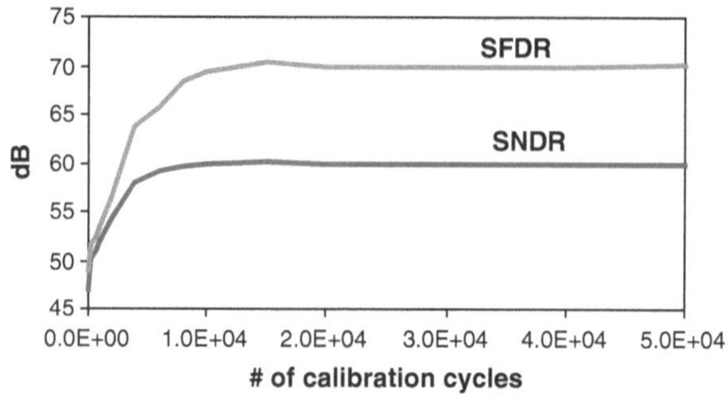

Fig. 6.26 ADC SNDR, SFDR improvement with # of calibration cycles

Table 6.1 Summary of ADC performance

Technology		1.8 V, 0.18 μm CMOS
Sampling rate (f_s)		45 MS/s
Input signal swing		1.3 V p-p
Area		3.57 mm^2
Power		81 mW
Number of calibration cycles		10^4 cycles (0.22 ms)
	Before calibration	After calibration
SNDR	46.9 dB	60 dB
SFDR	48.9 dB	70 dB
INL	+6.4/−6.1 LSB	+1.1/−1 LSB
DNL	+1.1/−0.4 LSB	+0.45/−0.4 LSB

6.7 Summary

A technique to rapidly measure and correct for missing codes introduced by a multi-bit first stage in a pipelined ADC was presented. Measured results from a prototype in 1.8 V, 0.18 μm CMOS show an improvement in INL from +6.1/−6.4 LSB to +1.1/−1 LSB, and SNDR/SFDR from 46.9/48.9 dB to 60.1/70 dB using the calibration approach of this work. Calibration was achieved in $\sim 10^4$ clock cycles. Table 6.1 summarizes key measurements of the work.

Chapter 7
A Power Scalable and Low Power Pipelined ADC

7.1 Overview

This chapter discusses the architecture of a reconfigurable ADC, which has a power which scales with sampling rate. A power scalable range over a large range of sampling rates is achieved without resorting to extensive current scaling, thus avoiding the problems of MOS transistors biased in weak inversion as described in Chapter 4. A general architecture for power scalable ADCs using a *Current Modulated Power Scale* (CMPS) approach is presented, where the application of CMPS to pipeline ADCs forms the focus of this chapter. Approaches to modulate current are presented, where a novel Rapid Power-On Opamp is shown to allow for CMPS to be used at high sampling rates in pipeline ADCs. The short on/off times of the Rapid Power-On Opamp are also shown to facilitate significant power reductions of opamp power in the MDACs of conventional pipeline ADCs, and more generally switched capacitor circuits, which have a clock phase that does not require a virtual ground. As such the MDAC stages are designed to power-off during the sampling phase, and optionally remain on during the sampling phase so that a measure of the power savings afforded by powering off the opamp during the sampling phase can be measured. Measurement results from a prototype fabricated in a 1.8 V, 0.18 μm CMOS process are given, where power scalability between 1 kS/s (15 μW) and 50 MS/s (35 mW) is achieved. Over the entire power scalable range the ADC achieved larger than 54 dB of SNDR. Furthermore, measurement results show that a 20–30% reduction in power can be achieved by powering off the opamps in a pipelined ADC for half a clock cycle.

7.2 Power Scalable Architecture

From Chapter 4 it is clear the many design problems associated with current scaling make it an undesirable power scalable approach for extended variations in f_s. Current scaling may be avoided however if the power scalable constraint is relaxed

I. Ahmed, *Pipelined ADC Design and Enhancement Techniques*,
Analog Circuits and Signal Processing,
DOI 10.1007/978-90-481-8652-5_7, © Springer Science+Business Media B.V. 2010

to average power rather than instantaneous power – i.e. $P_{avg}(f)$. Since the instantaneous power of ICs often varies significantly (due to digital circuitry), relaxing the power scalable constraint to average power is a valid compromise. The formula for power is altered to include average values as follows:

$$P_{avg} = \overline{P} \rightarrow \overline{P} = \overline{IV} \quad (7.1)$$

As mentioned in Section 4.3 voltage scaling is not a feasible approach, thus $\overline{V} = V_{DD}$,

$$\therefore \ \overline{P} = \overline{P(f)} = \overline{I}V_{DD} = \overline{I(f)}V_{DD} \quad (7.2)$$

Hence to obtain a power scalable average power, the average current must be frequency dependent. Although current scaling satisfies equation (7.2), alternative current scaling methods can be found which also satisfy equation (7.2), yet do no not impose the problems mentioned in Chapter 4. An alternative method can be derived by examining the ADC architecture, the settling requirements of the ADC, and the nature of digital circuits – which have their average power a function of frequency.

In digital circuits power is consumed only on output transitions where as described in Section 4.3 only enough power is consumed to set the digital output to the desired logic level. A characteristic of an ADC which may be exploited is although ADCs are predominantly analog, the final output is digital. Thus per output sample, the ADC only requires enough power to set the output to the logic levels representing the analog input. When an ADC operates at full speed, just enough analog power is supplied to allow the analog circuits to settle to the desired accuracy, and thus provide the correct digital output. If the sampling rate is decreased while maintaining constant bias currents (hence constant analog power), the opamp settling time remains unchanged but the figure of merit is reduced as the sampling rate decreases whereas the power remains approximately constant. As the ADC output is digital however, only enough power is required to charge/discharge the digital output to the correct logic levels, which as shown in Fig. 7.1, is achieved after $t_{ON} = t_{analog} + t_{digital}$ seconds.

Since the ADC is idle during t_{static}, if the digital output is latched from the ADC after t_{ON}, the analog portion of the ADC may be powered off after t_{ON} until the next sample is digitized as shown in Fig. 7.2.

Thus by powering off the analog portion of the ADC during t_{static} (which is related to the sampling frequency – the larger t_{OFF}, the lower the sampling rate), the ADC power can be made a function of effective sampling rate, i.e.

$$P_{avg} = P_{on}\frac{t_{on}}{t_{on} + t_{off}} = P_{on}\frac{t_{on}}{T_{effective}} = P_{on}t_{on}f_{effective} \quad (7.3)$$

where $f_{effective}$ is the effective sampling rate, and P_{on} the average power consumed by the ADC during t_{on}. The variation of average power with sampling rate is shown in Figs. 7.3 and 7.4:

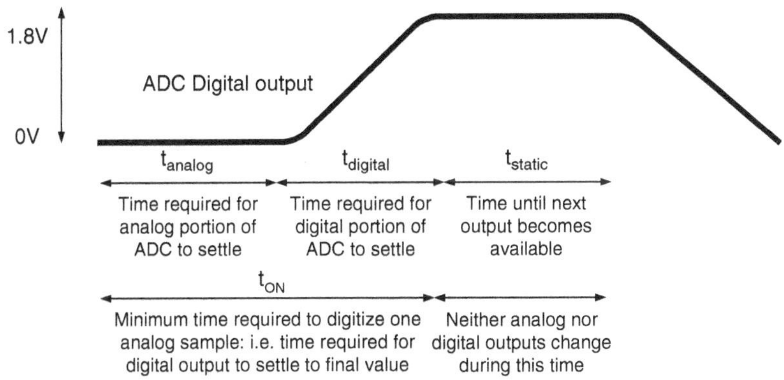

Fig. 7.1 Setup times for a nominal ADC

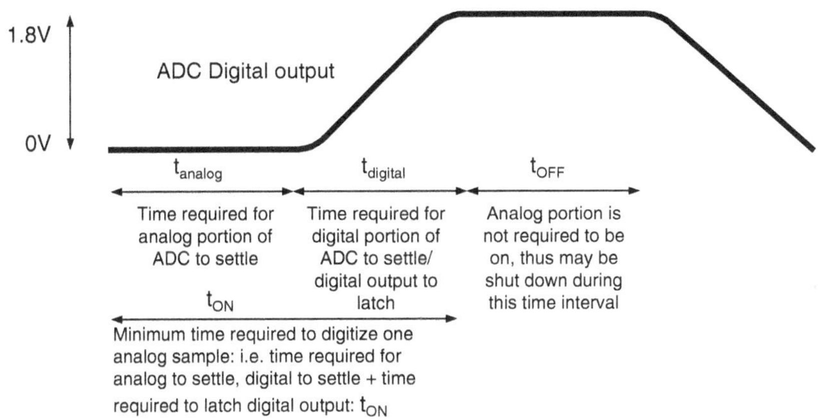

Fig. 7.2 Setup times for a current modulated ADC

Thus as the effective sampling rate decreases, ADC power is reduced by time averaging, and since P_{on} remains constant between sampling rates, the bias currents in the ADC, and thus the degree of channel inversion remain fixed. Hence the problems of poor device modeling, poor current matching, increased bias point sensitivity, and problems associated with IR drops of devices with low current biasing (hence weak inversion) are completely avoided. Since different power scaled sampling frequencies can be achieved while maintaining a constant on time (thus constant settling time), analog circuitry is minimally affected over frequency scaling. Thus the ADC characteristics for several sampling frequencies can be determined through simulating/testing only one sampling frequency, saving a significant amount of design, simulation, and test time over the current scaling method of Chapter 4. For example, the *Effective Number Of Bits* (ENOB) at $f_s = 10$ MHz is the same as $f_s = 1$ MHz, $f_s = 100$ kHz, $f_s = 10$ kHz, etc. as the bias current is unchanged between f_s – only the off time varies for different f_s.

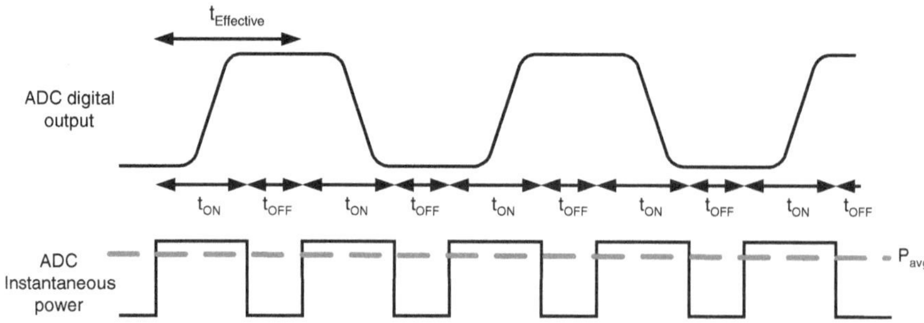

Fig. 7.3 Illustration of a high average power with modulated current

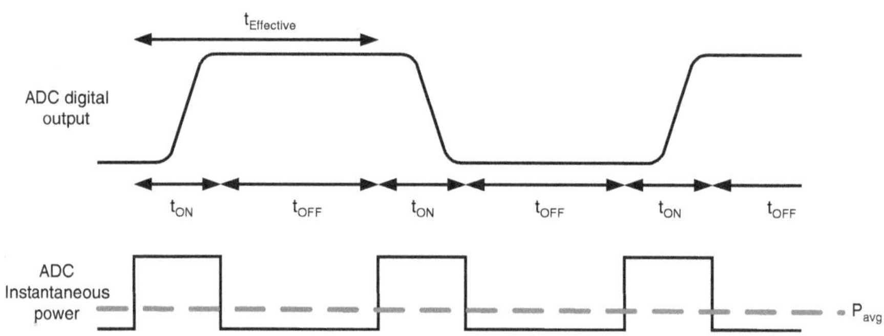

Fig. 7.4 Illustration of low average power with modulated current

It should be noted that if current scaling is used in addition to current modulation, the effective scalable range is the product of the two scalable ranges: (e.g.) If current scaling is used to scale the power between $f_s = 50$ MHz, and $f_s = 5$ MHz ($\times 10$ reduction in power), and the *Current Modulated Power Scale* (CMPS) technique allows for a $\times 100$ power scalable reduction in f_s, application of CMPS to the current scaled sampling rate of 5 MHz, allows for the sampling frequency to be reduced to 5 MHz/100 = 50 kHz without further reduction in bias currents. Thus, the scalable range is 10×100–1:1,000, although the bias currents are only scaled 1:10. As a result, the current modulation technique improves on the maximum achievable power scalable range possible with current scaling.

7.3 Current Modulated Power Scaling (CMPS)

From Fig. 7.2 the maximum sampling rate of an ADC that uses the CMPS technique is limited by the time to completely digitize one analog sample t_{ON} – the latency of the ADC. As CMPS powers down the analog portion during t_{OFF}, the ADC cannot have analog memory (through sample and holds) between output samples. Thus the speed advantage gained by pipelining stages, namely a sampling rate that is only $1/t_{stage\text{-}max}$ (where $t_{stage\text{-}max}$ is the maximum delay through a pipeline stage), is effectively removed in a pipeline ADC using CMPS. Since input samples that do not fully traverse the pipeline are erased when the ADC is powered off, the maximum sampling rate for a pipeline ADC using CMPS is $(Nt_{stage\text{-}max})^{-1}$, where N is the number of pipeline stages (each stage requires half a clock cycle to traverse). Although CMPS applied to pipeline ADCs limits the maximum sampling rate, the queue structure of the pipeline ADC lends itself to easy adaptation of the CMPS technique.

From equation (7.3), due to the large latency (hence large t_{ON}) of pipeline ADCs, powering off the ADC after a single sample is digitized leads to large average power consumption. Analog power of a pipeline ADC using CMPS however, can be reduced by powering the minimum number of pipeline stages per output sample. Consider the example of Fig. 7.5: the sampled inputs between 0.5 and 6.5 T do not fully traverse the pipeline ADC before the pipeline is powered off (i.e. these samples are not digitized to 10-bit accuracy), and thus the samples between 0.5 and 6.5 T may be ignored all together. By only powering the pipeline stages one at a time as the analog input sample traverses the pipeline, P_{ON} of the pipeline ADC using CMPS can be significantly reduced.

Therefore, by only powering a single stage at a time during digitization, the analog on power, P_{on}, can be reduced such that the average on power is the average

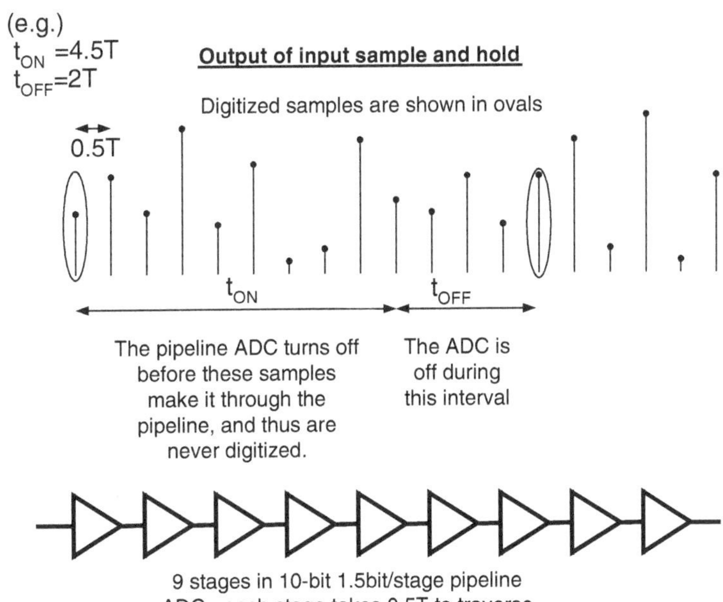

Fig. 7.5 Example illustrating the valid inputs to a pipelined ADC

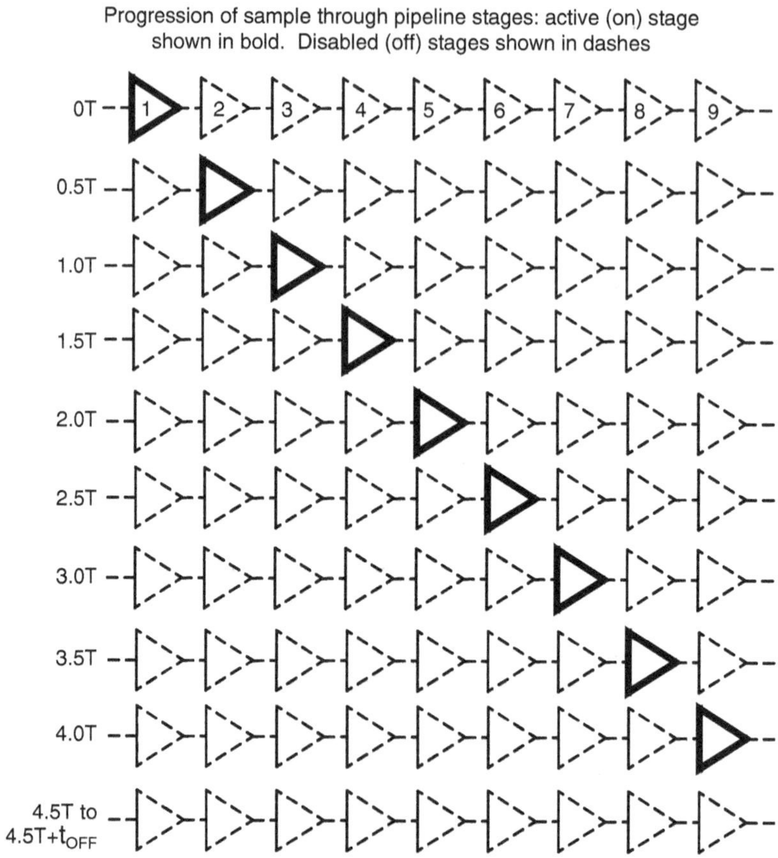

Fig. 7.6 On/off triggering sequence for a 10-bit pipeline ADC

power of a single stage. Figure 7.6 illustrates this concept, where the average power is shown to be:

$$P_{on-average} = average(P_{stage}) \tag{7.4}$$

In effect the pipeline ADC adapted to the CMPS technique operates as a Cyclic (Algorithmic) ADC, which uses successive pipeline stages, rather than reusing one pipeline stage over and over again. To adapt a high-speed pipeline ADC to use the CMPS technique to the following additional circuit blocks are required:

- Opamps that rapidly power on and off, with differential outputs that settle within one clock cycle after the opamp is powered on (need opamps similar to switched opamps [35–37], but at higher speeds to allow for high-speed ADC operation, and such that the power goes to zero during the off phase), to facilitate the pipeline stages powering on/off as in Fig. 7.6.

- A digital state machine to generate control signals to power on/off the various pipeline stages in a sequential manner (can be easily hand designed or synthesized. Shown in Section 7.7 to consume a small overhead power).

7.4 Current Switching Issues

From Sections 7.2–7.3, analog power scalability is shown to be possible by selectively powering on/off successive stages in a pipelined ADC. A consequence of powering analog blocks on/off is an increased power supply noise. When analog portions of the ADC are powered on/off the instantaneous current consumed by the power supplies changes a finite value in a very short time. As power supply noise is dominated by *Ldi/dt* noise (ground bounce) [26], a large change in current over a short time interval leads to undesirable fluctuations in V_{DD} and V_{SS}. Although a fully differential architecture minimizes the impact of supply noise on sampled and held signals, unavoidable asymmetries in the layout and/or signal swings lead to a finite manifestation of power supply noise on sampled signals. Since 10-bit accuracy levels require very low noise floors (less than 780 μV RMS noise for a 800 mV signal swing), it is crucial to maintain as constant a power supply as possible. Supply voltages can be held constant through an on chip regulator [38, 39], which provides constant supply voltages regardless of current variation. On chip regulators however are often not feasible due to limited power and area constraints, as such alternative methods are typically required. Most integrated circuits use passive circuits to minimize AC supply noise. By placing a passive RC filter with a pole near DC (where resistance is used to reduce the quality of the LC tank formed by the package parasitics), supply noise (which is primarily high frequency) can be adequately suppressed. A detailed analysis of power supply decoupling networks can be found in [26].

The RC filter of Fig. 7.7 was used to suppress power supply noise in this work, where due to a low series resistance MIM caps (~40 pF) were used for C_1, and due

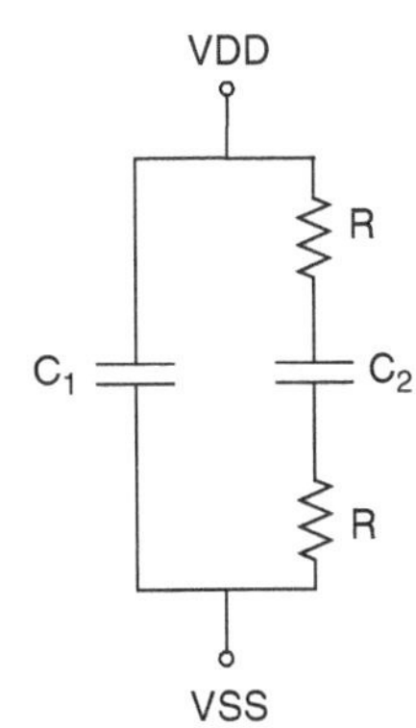

Fig. 7.7 Power supply noise decoupling circuit

to a higher series resistance [23], MOS capacitors (~150 pF) used for the combination of C_2 and R.

A significant advantage of applying CMPS to the pipeline architecture is on every clock transition there are at most only two opamps powering on/off (while one opamp powers off another powers on). Thus the pipeline latency serves to reduce *di*/*dt* as less current is switched per clock cycle.

7.5 Hybrid Power Scaling

As described in Section 7.3, the CMPS technique applied to a pipeline ADC limits the maximum sampling rate to $1/t_{ON} = 1/t_{latency}$, which for a 10-bit, 1.5-bit/stage pipeline ADC limits the maximum sampling rate to at best $f_{fullrate}/4.5$ (nine pipeline stages requiring half-a-clock cycle each to traverse). For example, applying the CMPS technique to a 50 MS/s pipeline ADC limits the maximum sampling rate to 50 MHz/4.5 = 11.1 MHz. Thus although the ADC is designed to run at 50 MS/s if operated as a conventional pipeline ADC, CMPS cannot be used to achieve scalable power between 11.1 and 50 MHz if 10-bit accuracy is desired as shown in Fig. 7.8 (assuming the ADC is operated as a conventional pipeline ADC for 11.1–50 MHz).

Since the range of sampling frequencies not covered by CMPS is small (only 1:4.5 in the example), a hybrid approach that uses both CMPS and current scaling can be used. If current scaling is used to provide a scalable power between the sampling rates not covered by the CMPS technique (i.e. 11.1–50 MHz) where the ADC operates as a conventional pipeline ADC, continuous power scalability results as shown in Fig. 7.9.

The problems of low current biasing (poor modeling, poor mismatch, increased bias sensitivity, and IR drops) are minimal with a hybrid approach as only a small current scaling range is required. Through careful design, the transistors can avoid operating in weak inversion over the narrow current scaling range. For example:

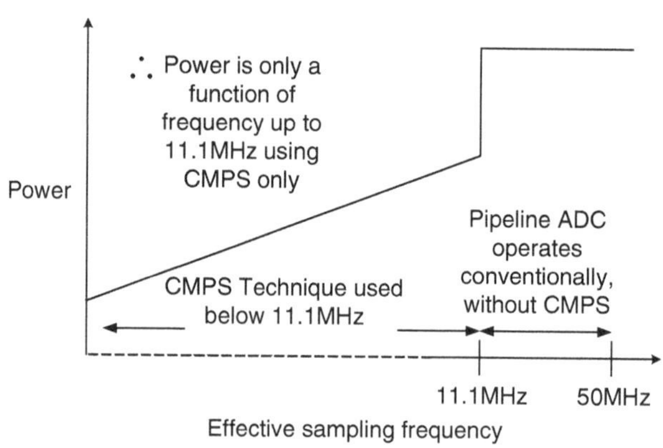

Fig. 7.8 CMPS limitations on power scalable frequency range

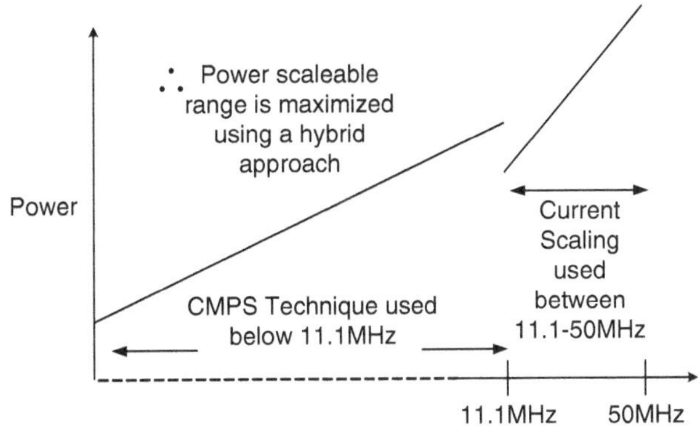

Fig. 7.9 Continuous power scalable range with hybrid power scaling

$V_{eff} \propto \sqrt{I_{DS}}$ to a first order, thus a variation of bias currents by $\times 10$ reduces V_{eff} by ~3.3. Thus if all current sources are designed to have a V_{eff} of 400 mV for higher speeds, V_{eff} is only reduced to ~120 mV for the lowest current-scaled sampling rate, which according to Section 4.4, places the device in moderate inversion.

A hybrid CMPS power scaling approach was used in the ADC of this chapter where current scaling was used to achieve scalable power for sampling rates not covered by CMPS. The hybrid CMPS approach was applied to a 10-bit 1.5-bit/stage pipeline ADC that was designed to have a maximum sampling rate of 50 MS/s.

7.6 Detailed Trigger Analysis

From Fig. 7.6, when using CMPS each stage opamp requires a trigger signal to power on/off the pipeline stages. In addition to the pipeline stages, other analog blocks must also be powered on/off, as (7.3) is based on the entire analog portion powering off during t_{OFF}. If certain circuit blocks are always on (i.e. are not powered off during t_{off}, and thus have a static power), the power scale formula of (7.3) is modified to

$$P_{avg} = P_{on} t_{on} f_{effective} + P_{static} \tag{7.5}$$

$$\therefore \lim_{f \to 0} P_{avg} = P_{static} \tag{7.6}$$

Thus as the sampling rate decreases, the power becomes less dependent on frequency and ultimately limited by P_{static}. Hence it is essential to minimize the number of blocks that are always powered on, so as to maximize the power scalable range. A system level diagram of the ADC of this work showing each major block is shown in Fig. 7.10.

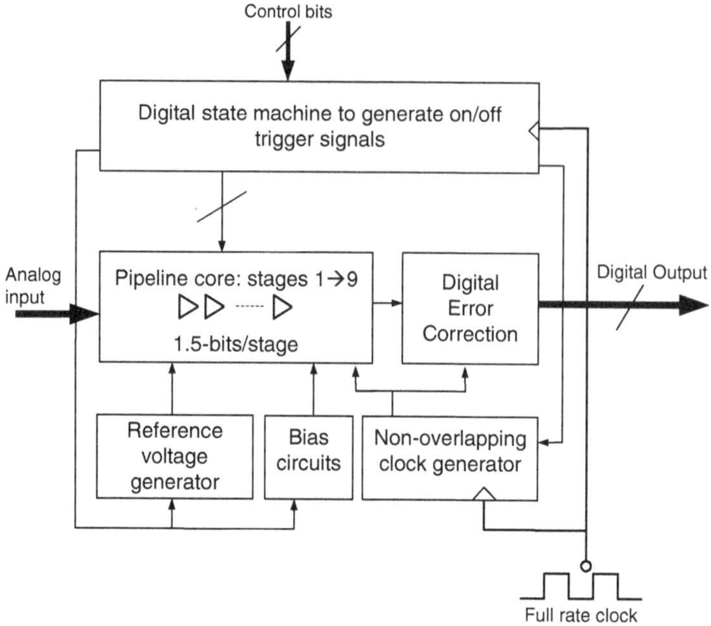

Fig. 7.10 Major sub-blocks in a 1.5 bit/stage pipeline ADC using CMPS

In addition to the pipeline core (MDACs + stage ADCs), a power on/off scheme is required for the bias circuits, clock generator, reference generator, and digital error correction.

The digital state machine never powers down, as it is required to generate the on/off trigger signals for each block, thus contributes to P_{static}. As will be shown in Section 7.7 however, the average power of the digital state machine can be made low, thus facilitating a large power scalable range.

As each pipeline stage powers on/off according to Fig. 7.6, the bias circuit for each stage may also power on/off for the same time interval. However as bias nodes are often loaded with large capacitances, and set with low currents, the time required to power on/off a bias circuit to a minimum settling accuracy can be very large (relative to the on/off time of a pipeline stage – Fig. 7.12). Thus with the setup shown in Fig. 7.11 the bias circuit of stage X cannot be synchronously triggered with stage X, as the bias voltages do not settle quickly enough for proper operation of stage X.

By powering on the bias circuit of stage X before stage X powers on however, the bias circuit of stage X can settle to a minimum accuracy before stage X is powered on as shown in Fig. 7.12.

The additional setup time for the bias circuit however increases the minimum on time of the ADC, t_{ON}, as the bias for stage 1 is required before stage 1 becomes active.

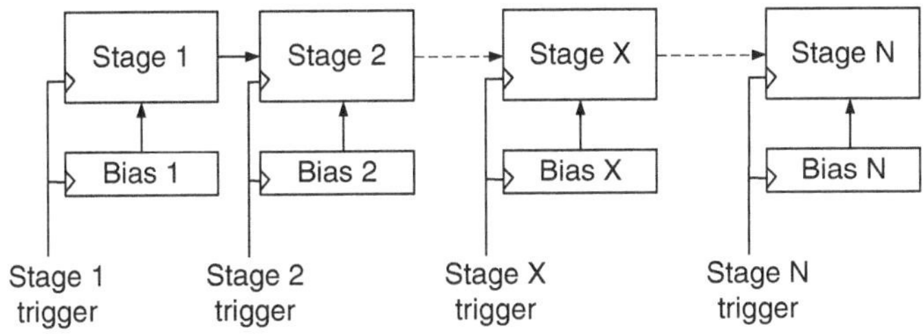

Fig. 7.11 One to one stage biasing arrangement

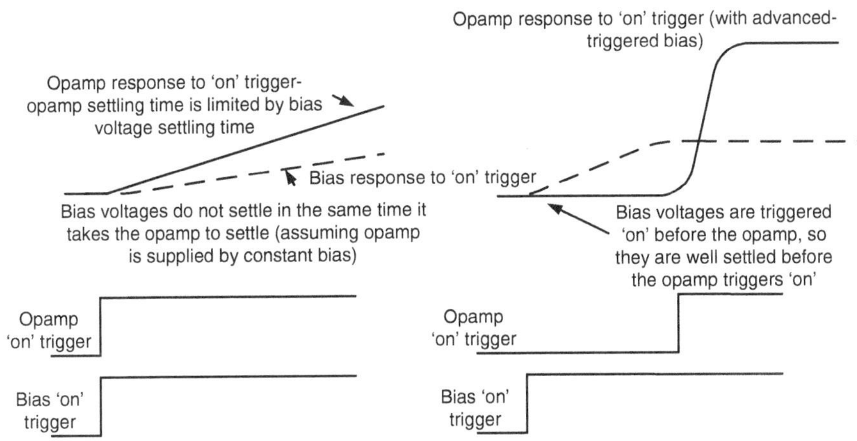

Fig. 7.12 Illustration of different bias circuit on/off techniques

In addition to the bias circuit, the ADC requires a clock generator and reference voltage generator. Although the digital state machine always requires a clock, the non-overlapping clocks required by the pipeline stages can be powered off during t_{OFF}. Turning off the clock generator during t_{OFF} can save a large amount of power as the non-overlapping clock generator drives a large capacitive load, hence consumes a large average power when on. To ensure the reference voltages have settled to a minimum accuracy, and the non-overlapping clock generator is fully powered on, the two blocks can be powered on a few clock cycles before the trigger to the first stage. As the digital error correction block only operates while the pipeline core outputs transitioning bits (i.e. is inherently a power scalable block), an on/off trigger is unnecessary for the block.

Some analog blocks however cannot be powered on/off in a reasonable amount of time. One such block is a current mirror that receives an off-chip bias current as shown in Fig. 7.13 (i.e. constant current bias circuit – used to set on-chip bias currents).

As node V_A in Fig. 7.13 has a large time constant, the settling time after an off to on transition can be excessive due to a large RC time constant/slew time from large

Fig. 7.13 A power on/off scheme for current mirror biased by off chip resistor

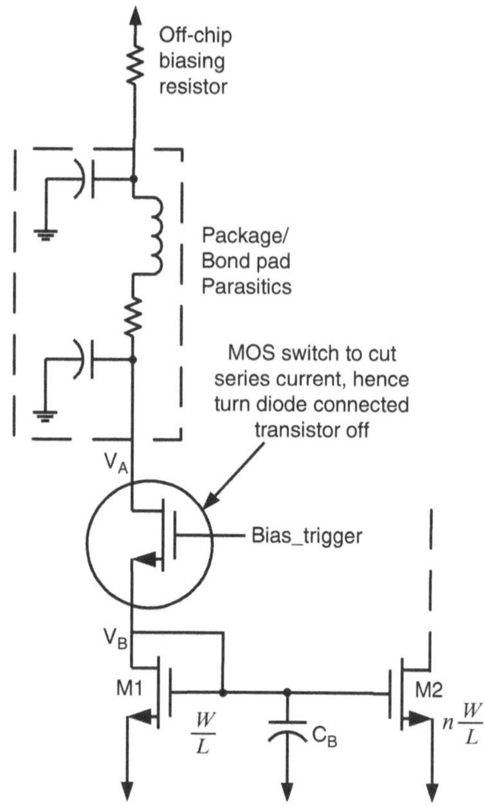

bond pad capacitances and off chip resistance. To maintain stable on chip biases voltages, the diode-connected transistor must always remain on – which contributes to P_{static} thus limiting the maximum power scalable range. The static power can be minimized however by supplying only a small off chip current, where larger on chip currents can be generated on chip by appropriate mirror transistor sizing (e.g. M2 in Fig. 7.13 has n times the drain-source current of M1) of current mirrors which can be powered on/off. The triggers signals for each major ADC block are shown in detail in Fig. 7.16. To simplify the bias on/off triggering, all bias circuits are activated on the same clock edge. A tradeoff of activating all bias circuits on the same clock edge is average power is increased. Future work could investigate optimizing the bias circuit timing such that a minimal power average is consumed. To save power and area, only three bias circuits have been designed for the ADC, such that one bias circuit serves three stages as shown in Fig. 7.14, where each bias circuit receives a reference current from a master current source set by an off chip current.

The power of each bias circuit can be modulated by a series current switch (MS in Fig. 7.15), which cuts the DC current path between supplies.

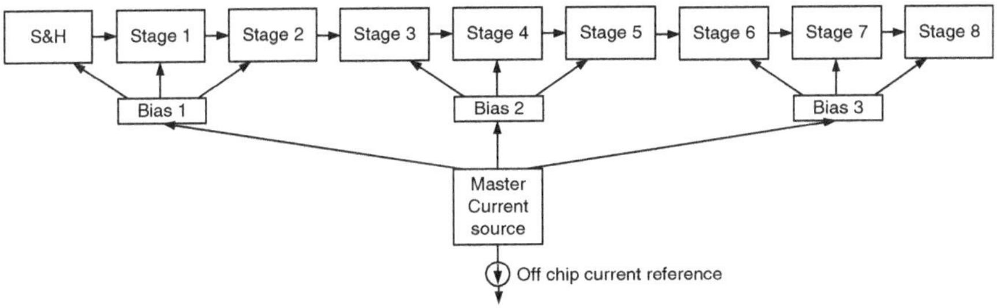

Fig. 7.14 Bias current routing for ADC

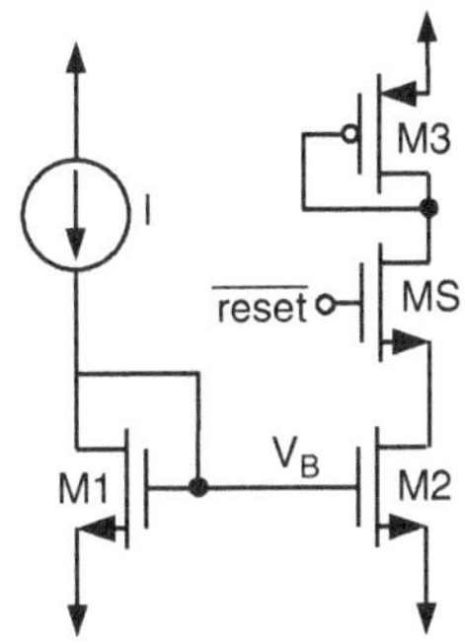

Fig. 7.15 Current switch 'MS' modulates bias circuit power

7.7 Design of the Digital State Machine

A digital state machine is required to generate the control signals Fig. 7.16 in uniform time intervals. By counting the number of clock edges of a full rate clock, precise timing of the control signals can be achieved. (For example): the on/off trigger for stage one is enabled when the counter counts to N, the on/off trigger for stage two is enabled when the counter reaches N + 1, etc. Similarly the other control signals can be generated when the counter reaches a pre-programmed value. The effective sampling rate of the ADC can be controlled by resetting the counter after the counter has counted K clock cycles of the full rate clock. The effective sampling rate (the rate upon which the power scales with) is thus given by

$$f_{effective} = \frac{f_{fullrate}}{K}, \text{ i.e. } T_{effective} = KT_{fullrate} \qquad (7.7)$$

Thus the effective sampling rate is not set by adjusting the off chip clock (full rate clock of Fig. 7.10), rather is digitally controlled by adjusting the value of 'K' in the state machine. The state machine was manually designed with a programmability, via serially loaded control bits, that allows for the adjustment of (1) The effective sampling rate (i.e. the value of K). (2) The lead setup time for the bias

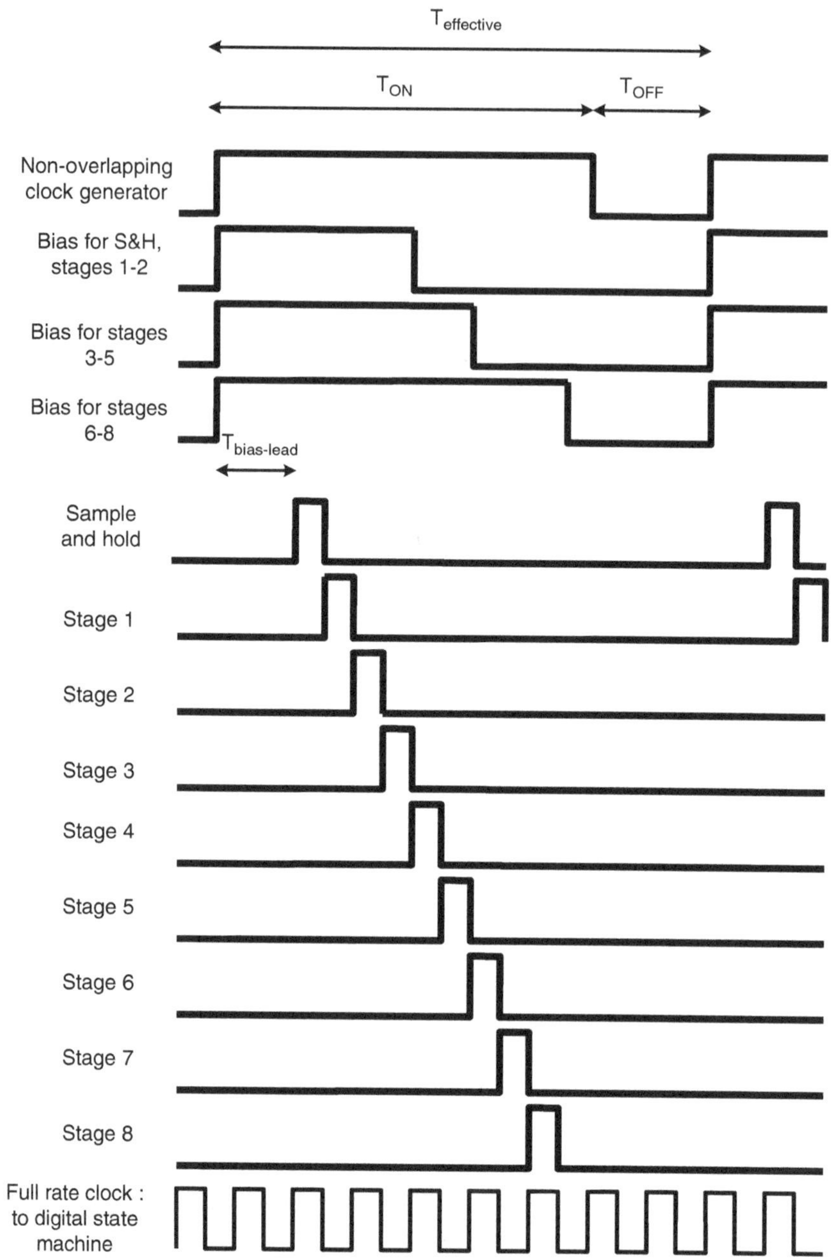

Fig. 7.16 Detailed triggering diagram for pipeline ADC using CMPS (stage 9 does not require a power on/off trigger as it only consists of dynamic comparators)

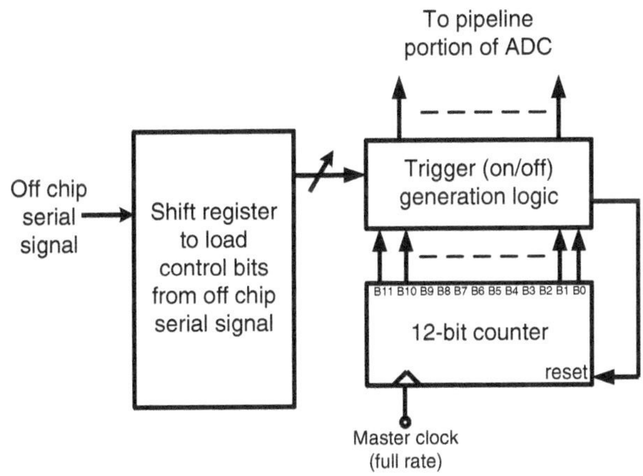

Fig. 7.17 System level diagram of on/off trigger generating digital state machine

Table 7.1 Variation of digital state machine power with clock frequency

Frequency (MHz)	Power (μW)
1	~9.2
10	~92
100	~920

circuits, $t_{bias\text{-}lead}$ (as the exact bias lead time required is difficult to determine through simulation due to the many parasitic capacitors on the bias nodes in the layout). The counter used in the digital state machine was a synchronous 12-bit counter – limiting the lowest clock speed to $1/(2^{12}-1) = 1/4{,}095$ the full-rate clock. For example, 50 MHz/4,095 = 12 kHz if 50 MHz provided to the state machine, 1 MHz/4,095 = 240 Hz if 1 MHz provided to the state machine. A system level diagram illustrating the state machine is shown in Fig. 7.17.

The state machine power consumption at various full rate speeds is shown in Table 7.1. Since the state machine is always on, the power of the ADC is ultimately limited to at least the power of the state machine. However as the digital power consumed by the state machine is a function of the full rate clock in Fig. 7.17, if the clock rate is reduced, the state machine power can be reduced. Since the settling time of the MDACs in the pipeline is related to the period of the full-rate clock, for slower clocks supplied to the state machine, the bias currents supplied to the MDACs can be reduced to maintain an optimal figure-of-merit at the penalty of less inverted MOS transistor channels.

It should be noted that if the digital state machine were synthesized using commercial logic gates, the power could be significantly reduced (at least ×2–5), maximizing the power scalable range (A conservative digital design approach was taken to ensure functionality, where the state machine was manually designed).

7.8 Rapid Power-On Opamps

As described in Section 7.3, the CMPS technique requires an opamp that powers on and off in a short time interval (so as to maximize the sampling rate). Switched opamps, are similar in functionality to the desired Rapid Power-On Opamp. Switched opamps are often used in low voltage applications [35–37], and are typically limited to less than 25 MS/s. To achieve rapid power-on, switched opamps typically do not completely power-off. The majority of switched opamps operate by switching bias voltages, or by series current switching. Sections 7.8.1–7.8.4 provide a brief overview of the implementation and design issues associated with each approach, as well as the Rapid Power On approach of this work.

7.8.1 Conventional Approach: Switched Bias Opamp

Switching the bias voltages of current source transistors in opamps modulates the opamp current and hence power. The switched bias approach to power on/off the opamp however has several design issues that lead to long power on/off times. Consider the schematic of Fig. 7.18, which shows the use of bias switching to achieve power on/off operation, where M1 is a current source transistor used to provide bias current in an opamp which is to be powered on/off.

In the topology of Fig. 7.18, when the current source powers-on, the time required for the bias voltage (V_{B2}) to settle to the desired bias voltage (V_{B1}) is limited by the RC time constant of the switch network. As bias voltages are required to stay constant during nominal operation, large decoupling capacitors C_1, and C_2, are typically connected to V_{B1} and V_{B2} respectively. Furthermore as mirror transistors tend to be large in area (to minimize mismatch, or introduce a current gain), large parasitic capacitances exist on the bias nodes. When φ switches from low to

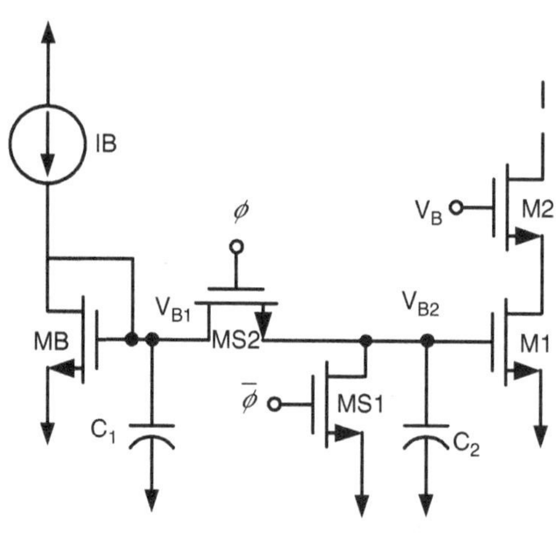

Fig. 7.18 Switched bias approach to turn the current source M1 on/off

high in Fig. 7.18, a discharged C_2 is placed in parallel with a charged C_1, temporarily causing the voltage at V_{B1} to dip to $C_1/(C_1 + C_2)V_{B1\text{-steady state}}$ due to charge sharing. For V_{B1} and V_{B2} to settle to $V_{B1\text{-steady state}}$, V_{B1} slews to the correct voltage according to the total capacitance of $C_1 + C_2$, and the difference in current of IB and that drawn by MB. The slewing current is small, and $C_1 + C_2$ is large, thus the slew time can be very long (consider that for this work when CMPS is enabled at the highest operating speed the opamps have less than 0.5*(1/50 MHz) = 10 ns to settle to at least 10-bit accuracy). Alternatively C_1 can be made very large (which consumes area) to minimize the amount V_{B1} dips when shorted to V_{B2} through MS2. However, if C_1 is made very large (e.g. 10's of pF), the slew rate at V_{B1} becomes very low, and thus although the dip in V_{B1} becomes small, V_{B1} never settles as it continually slews duringφ. With mismatches between positive and negative halves, and asymmetric parasitic capacitances in a fully differential opamp, if the bias voltages to it slew, the differential output of the opamp also never settles, thus affecting the settling accuracy of the switched capacitor circuit which is of paramount importance in a 10b pipeline ADC.

7.8.2 Rapid Power-On Opamps Used in This Work

The settling times of master bias voltages (the bias voltage generated by diode-connected transistors) have been shown to limit the minimum power-on time of opamps when using a switched bias/bias reset scheme to modulate opamp power. If however the master bias voltages could be held constant while the opamp is powered on/off, the opamp power on/off time could be significantly reduced. A solution to this problem is achieved by switching a copy of the bias voltage rather than the bias voltage itself. Using negative feedback, it is possible to copy a bias voltage from one node to another as shown in Fig. 7.19.

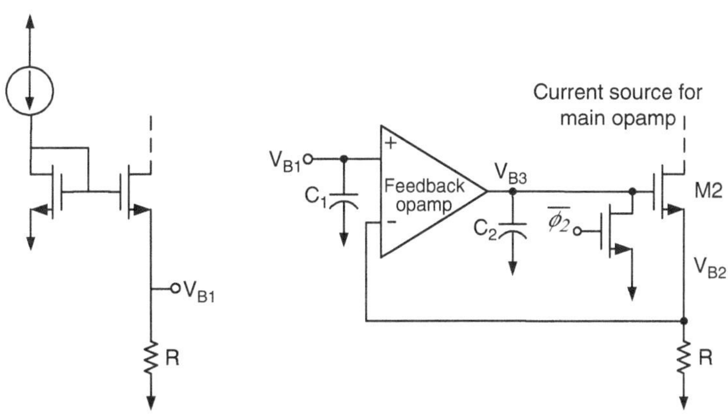

Fig. 7.19 Feedback based bias switching

By virtue of negative feedback, $V_{B2} \approx V_{B1}$, and due to the large input impedance of the opamp, V_{B2}, and V_{B3} are well isolated from V_{B1} (and vice-versa). Thus switching V_{B3} to modulate the current of M2 minimally disturbs V_{B1}. Furthermore, as V_{B3} is isolated from V_{B1}, the RC time constant/slewing time at V_{B3} is minimal, as V_{B3} does not share any parasitic and decoupling capacitance with other bias nodes. Also charge sharing effects from switching at node V_{B3} do not affect the bias source (V_{B1}), thus avoiding excessive settling times as described in Section 7.8.1. Thus, V_{B3} can be switched quickly without disturbing V_{B1}.

For the CMPS technique however, the entire analog core must power off for a fixed time interval, thus the feedback opamp must also power-off. On the surface using an opamp to shorten the on/off time of another opamp seems like a cyclical argument, however by exploiting the different performance requirements of the main opamp (i.e. the opamps used in the MDACs) and the feedback opamp (the opamp used to shorten the power on time of the main opamp), rapid power-on can be achieved without disturbing the bias voltages. Consider the switching scheme of Fig. 7.20 to power on/off an opamp, where M2 is a current source which provides bias currents in an opamp:

The series switch approach shown in Fig. 7.20 avoids perturbing the bias voltages, as the switch transistor M3 switches current rather than the bias voltage. Current switching is often used for low power switched opamps [37]. Current switching only depends on the time required to toggle the switch transistor from cut-off to triode, thus is very fast and much faster than bias switching approaches, as the gate of the switch transistor requires no decoupling capacitance, thus has a small RC time constant. Current switching however reduces the available signal swing due to an IR drop across the switch transistor when in triode, which can be significant. (For example) a low triode resistance for a switch is 100 Ω. If the current to be switched is 1 mA, the IR drop is 100 mV, which is excessive for a 1.8 V supply, where signal swings tend to be 500–1,000 mV (single-ended). Thus the current switch method is not preferable in opamps where large bias currents are used and/or a large signal swing is required (i.e. the main MDAC opamp). Series current switches, which are designed with the intention of operating in the triode region, can also shift to the active region as the output signal swing in an opamp

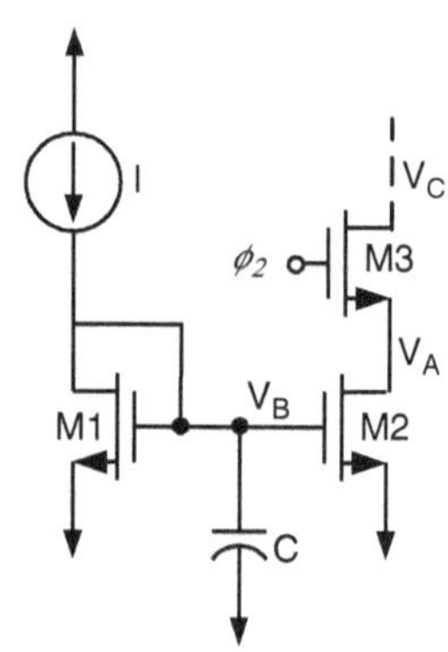

Fig. 7.20 Series switching to turn M2 on/off

moves between minima and maxima. For differential opamps where when one output is at a maximum the other is at a minimum, series current switches in the output stage of an opamp can lead to the switch being in active for only one half of the circuit, which leads to a degradation of circuit symmetry and thus power supply noise rejection.

The feedback opamps of Fig. 7.19 however, drive a much smaller load (in comparison to the main opamp) thus have lower bias currents, and only require a small output swing (since the output need only include the variation of the bias voltage – which is very small), hence can tolerate a reduction in available signal swing. Furthermore since the feedback opamps are single ended, they do not require differential symmetry. Thus current switching can be applied to the feedback opamps as shown in Fig. 7.21, where the trigger signal to power on/off the feedback opamp are applied to node 'reset'.

The on/off time of the feedback based bias approach can be adjusted by minimizing the RC time constant at node V_{B3} in Fig. 7.19, and/or by increasing the bandwidth of the feedback opamp. (cf. the time constant of a closed loop opamp is g_m/C_{load}, hence the settling time can be controlled by exchanging power for speed). Thus by combining different power on/off techniques – a feedback based bias in the main opamp and current switching in the feedback opamp, a very small power on/off time can be achieved.

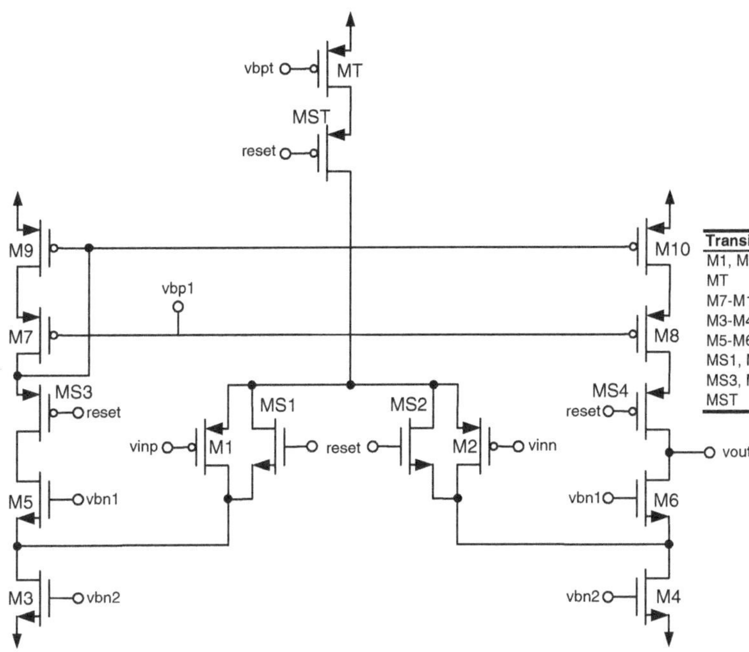

Transistor	W/L
M1, M2	5/0.24 x 3
MT	5/0.24 x 18
M7-M10	5/0.24 x 9
M3-M4	3/0.24 x 6
M5-M6	3/0.24 x 3
MS1, MS2	3/0.18 x 5
MS3, MS4	5/0.18 x 9
MST	5/0.18 x 10

Fig. 7.21 Feedback opamp with current switching

7.8.3 Benefits of Feedback Based Biasing: Increased Output Resistance

A drawback of the feedback based biasing scheme of Fig. 7.19 is the requirement of additional opamps hence additional power consumption. However, feedback based biasing serves to increase the output resistance of the current source transistors as shown in Fig. 7.22:

Feedback based biasing can be arranged in a gain-boosting configuration [21], such that a large gain can be achieved in the main opamp using only a single stage architecture. From Section 3.4, a large DC gain is a necessary requirement for the opamps in a pipeline ADC, thus using a feedback based biasing approach is doubly beneficial. By combining the feedback based switching approach with a folded cascode opamp, a gain-boosted single stage opamp with short power on/off times, and large DC gain results as shown in Fig. 7.24 (where signal *reset* powers on/off the opamp, and SFBO is the *Switched Feedback Bias Opamp* of Fig. 7.21 for NMOS gain boosting, and the opamp of Fig. 7.23 for PMOS gain boosting.):

As the feedback opamps provide increased opamp gain, the additional power required to power the feedback opamps is minimal as multiple opamp stages (thus more complicated compensation schemes) to achieve large gain are avoided.

A folded cascode architecture is beneficial as it allows the opamp input common mode to include ground. With an input common mode near ground, it is possible to use the current switch technique on the tail current transistors as a large IR drop can be tolerated across transistor MST (in Fig. 7.24) while maintaining the input differential pair in saturation. Transistors MS3 and MS4 are used in parallel with the feedback opamps to shorten the power-off times of M5 and M6. MS1 and MS2 are used to set node V_X to a defined voltage during the power-off state. MS5 & MS6 quicken the opamp-reset time, pulling the outputs to V_{DD} as required by the common-mode feedback circuit (Section 7.9). The opamp was biased such that the differential peak-to-peak output swing was 1.6 V with the maximum bias current.

In Figs. 7.25 and 7.26 the transient responses of two approaches to power-on the opamp are compared. The simulation shows the transient response of an MDAC using the feedback based approach of Fig. 7.24, and an MDAC using a folded cascode opamp where the bias voltages to the opamp are switched as shown in

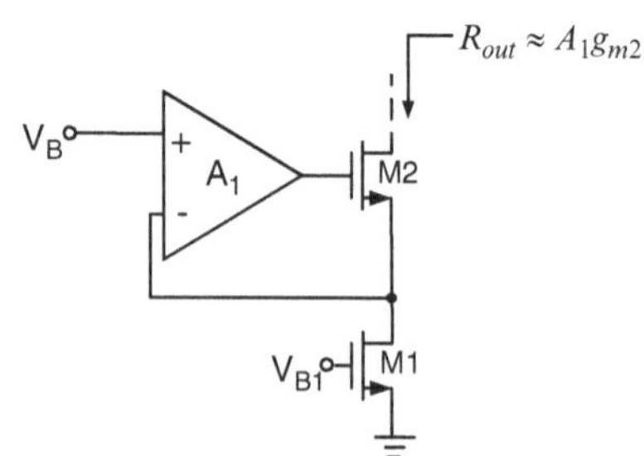

Fig. 7.22 Increased output impedance using feedback based biasing

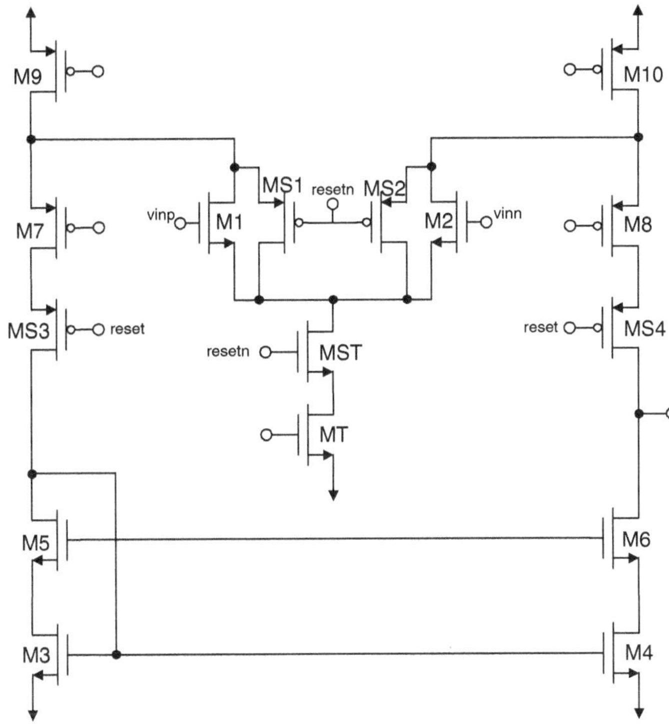

Transistor	W/L
M1, M2	3/0.24 x 12
MT	3/0.24 x 12
M7-M8	5/0.24 x 18
M9-M10	5/0.24 x 36
M3-M6	3/0.24 x 6
MS1, MS2	5/0.18 x 6
MS3, MS4	5/0.18 x 18
MST	3/0.18 x 12

Fig. 7.23 Switched PMOS gain boosting opamp

Fig. 7.18, with multiple opamps sharing the same master bias. From the figures it is clear the feedback based bias approach provides fast on/off times facilitating fast sampling rates.

7.8.4 Opamp Specification/Characterization

As described in Sections 3.4 and 3.5, the opamps for each stage in the ADC pipeline require a minimum DC gain and bandwidth to achieve 10-bit accuracy. Ideally each stage would be uniquely designed such that the required gain/bandwidth specifications are just met to minimize area and power. However, scaling each stage requires the design and layout of eight unique opamps for a 1.5-bit/stage 10-bit pipeline ADC. As the goals of this prototype more favor proof-of-concept over absolute performance specifications, stage opamps were scaled in groups rather than individually as shown in Fig. 7.27.

To achieve 10-bit accuracy at a sampling rate of 50 MS/s, the stage opamps were designed with the DC-gain and unity gain bandwidths shown in Table 7.2.

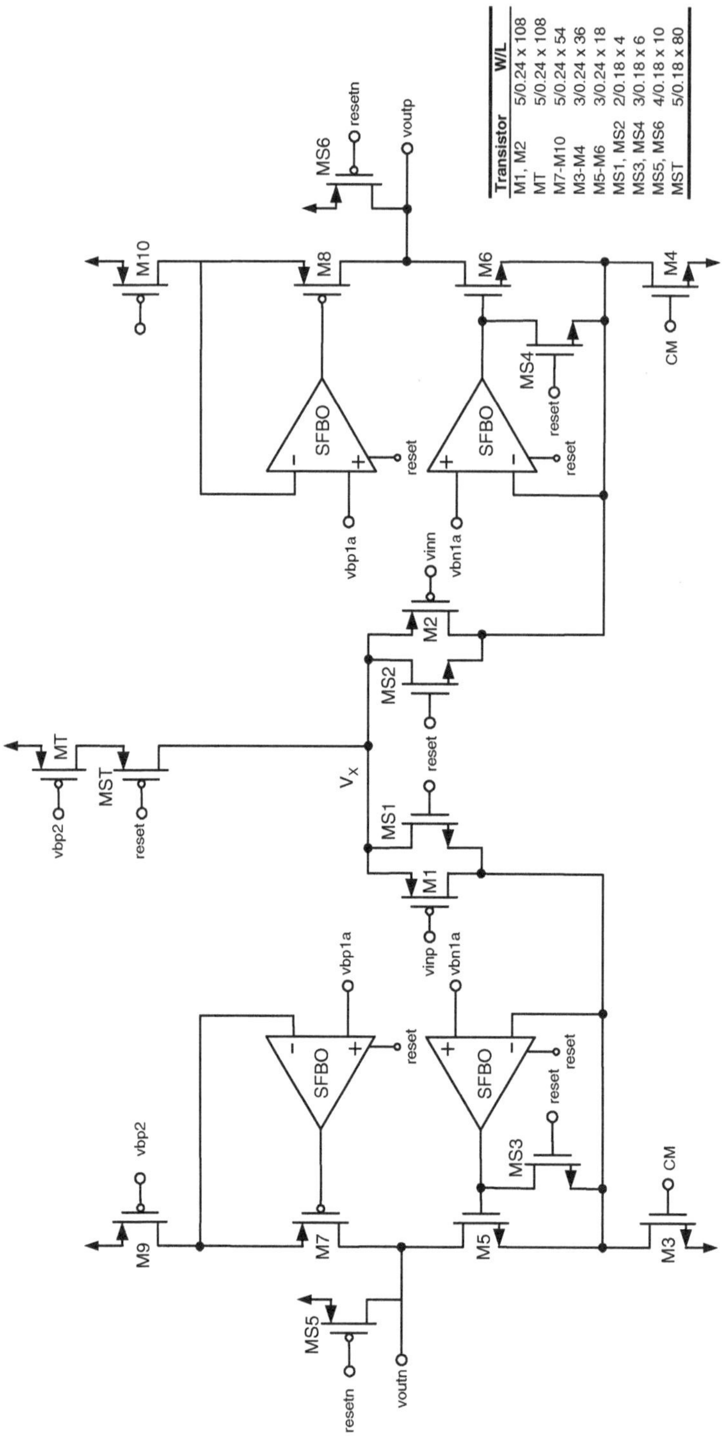

Transistor	W/L
M1, M2	5/0.24 x 108
MT	5/0.24 x 108
M7-M10	5/0.24 x 54
M3-M4	3/0.24 x 36
M5-M6	3/0.24 x 18
MS1, MS2	2/0.18 x 4
MS3, MS4	3/0.18 x 6
MS5, MS6	4/0.18 x 10
MST	5/0.18 x 80

Fig. 7.24 High gain feedback based switched opamp

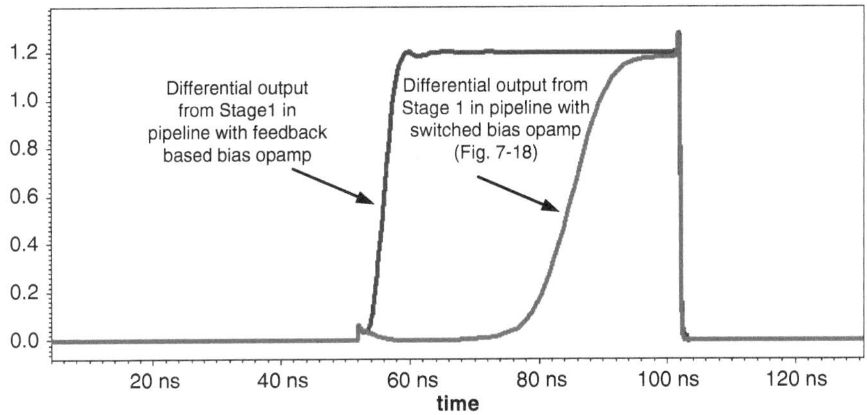

Fig. 7.25 SPICE simulation comparing different switching approaches

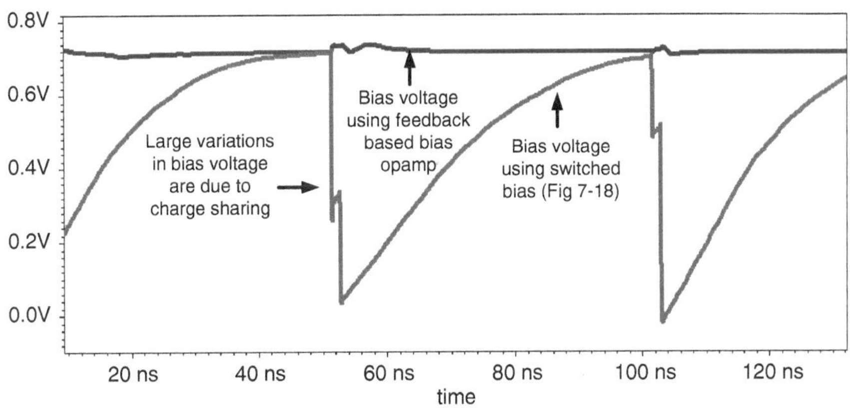

Fig. 7.26 SPICE simulation showing impact of switching architecture on bias voltages

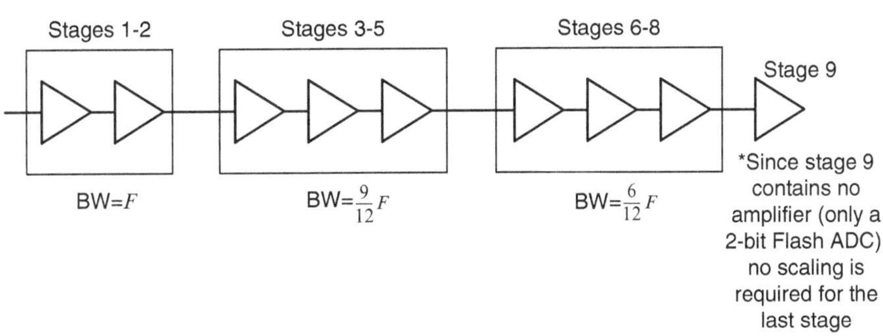

Fig. 7.27 Stage grouping for scaling

Table 7.2 MDAC Opamp DC gain and bandwidth for 50 MS/s operation

Stages	DC gain (dB)	Maximum unity gain (MHz)	Desired relative unity gain	Phase margin (degree)	MDAC sampling/ feedback capacitor	Effective opamp load (pF)	Opamp power (mW)
1, 2	105	563	f	75	Os.5	1.4	7.8
3, 4, 5	73	515	$\frac{9}{12}f$	62	0.1	0.36	3.3
6, 7,8	52	297	$\frac{6}{12}f$	80	0.05	0.23	1.1

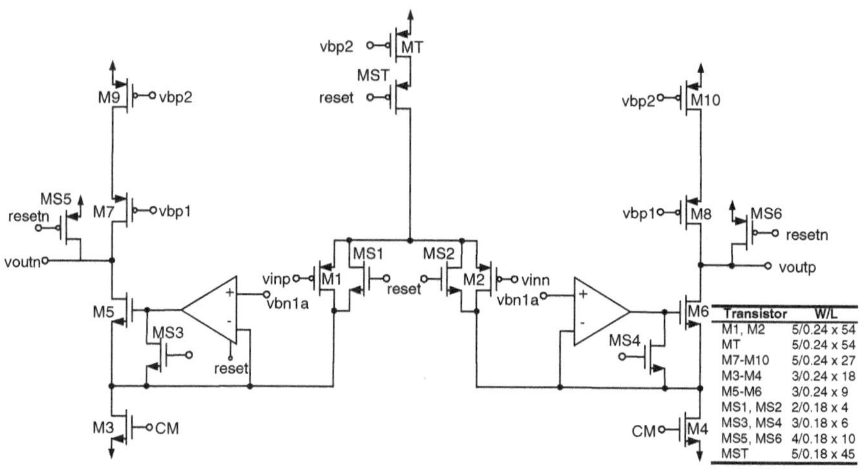

Fig. 7.28 Opamp for stages 3–5

Reduced opamp gain for latter stages in the pipeline was achieved by removing the feedback based biasing opamps as shown in Figs. 7.28 and 7.29. The opamps for stages 6–8 use series current switching, as the opamps in stages 6–8 have small bias currents, and require less SNR, thus can tolerate a smaller signal swing.

As mentioned in Section 7.8.3, a benefit of the feedback based bias/gain boosted architecture is the avoidance of complicated compensation structures. As the opamps have essentially a single stage architecture, simple load compensation can be used to achieve a minimum phase margin. An added benefit of load compensation is any additional parasitic capacitance at the opamp's output that is not accounted for in simulation, but manifests in the layout process only serves to enhance the phase margin [23].

The bandwidths of the switched feedback based bias opamps were tuned through simulation such that the closed loop response is stable and short settling times result. The feedback based bias opamps setting the bias voltage of the NMOS transistors were biased with one sixth the bias current of the main opamp and the PMOS feedback opamp biased with one third the bias current of the main opamp.

A Monte Carlo Analysis was performed in SPICE to determine opamp bandwidth variation as bias current is reduced. The relative variation (3σ/mean) bandwidth as

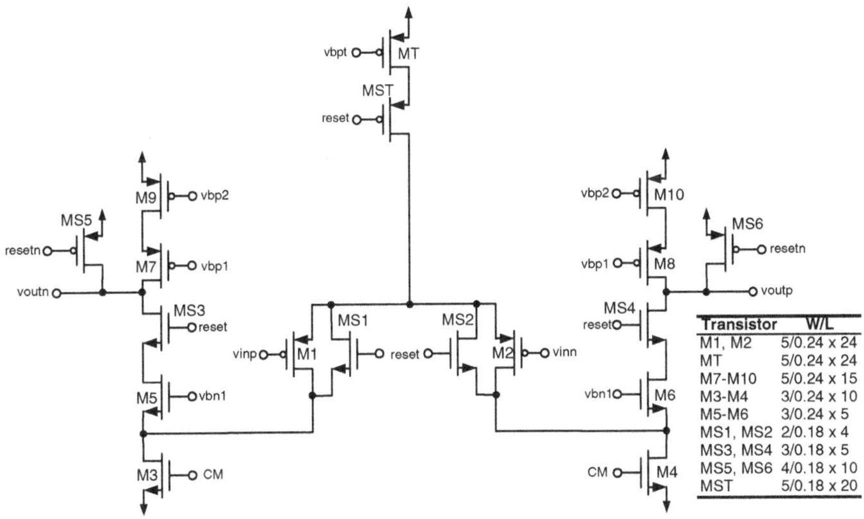

Transistor	W/L
M1, M2	5/0.24 x 24
MT	5/0.24 x 24
M7-M10	5/0.24 x 15
M3-M4	3/0.24 x 10
M5-M6	3/0.24 x 5
MS1, MS2	2/0.18 x 4
MS3, MS4	3/0.18 x 5
MS5, MS6	4/0.18 x 10
MST	5/0.18 x 20

Fig. 7.29 Opamp for stages 6–8

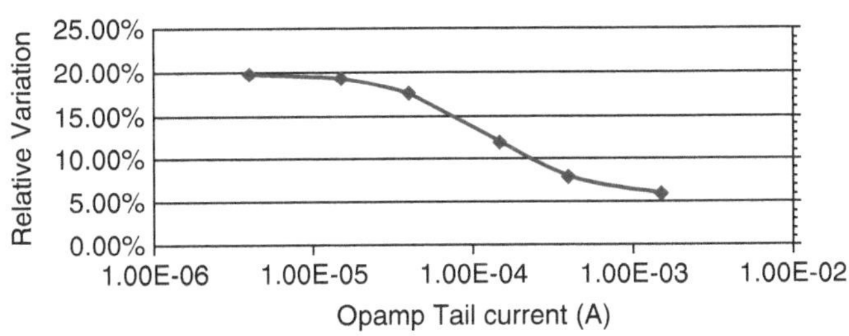

Fig. 7.30 Relative variation (3σ/mean) of opamp bandwidth versus tail current of opamp in Fig. 7.24

the opamp tail current is decreased is shown in Fig. 7.30. The larger variation as current is decreased verifies the predicted poorer matching as the opamp is driven deeper into the weak inversion region of operation.

7.9 Common Mode Feed Back (CMFB) for Rapid Power-On Opamp

One of the requirements of CMPS (according to the triggering diagram of Fig. 7.16) is to have the MDAC output fully settled within one clock cycle after the stage trigger signal is enabled. Typically switched capacitor circuits use a switched capacitor CMFB (Fig. 7.31) circuit which takes several clock cycles to generate the correct common-mode in the output [10] – clearly not feasible for a pipeline ADC using CMPS.

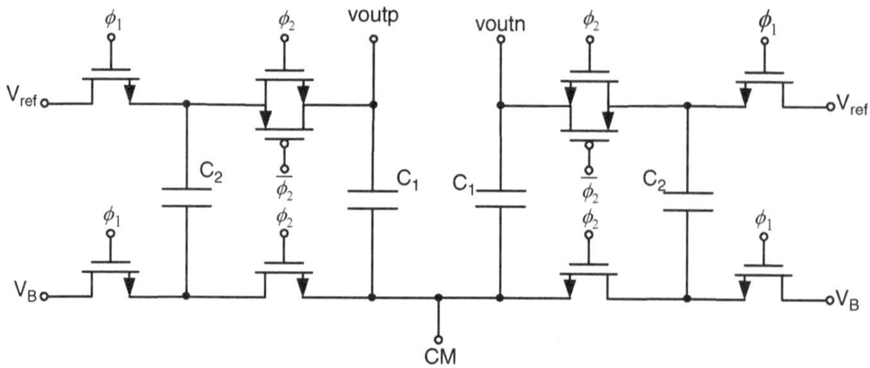

Fig. 7.31 Conventional passive switched capacitor CMFB circuit

Fig. 7.32 Passive switched capacitor circuit for switched opamps

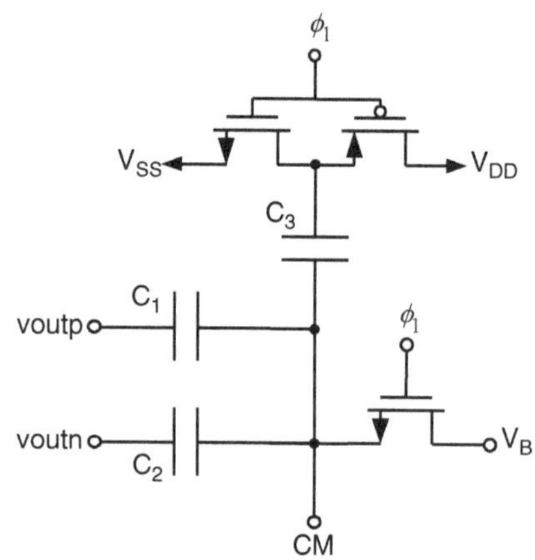

The CMFB circuits of switched opamps however, can be used as switched opamps have differential outputs, which settle within one clock cycle after a the opamp is powered-on. CMFB approaches for switched opamps fall into passive [37] and active [36, 41] categories. Active CMFB approaches have the advantage of settling to a common mode voltage defined by a known reference voltage, but have the penalty of additional power consumption to power the active circuitry. Passive approaches achieve the desired common-mode level by charge conservation, and thus have a common mode voltage that is sensitive to parasitic capacitances of the CMFB circuit. Passive CMFB however has the advantage of only consuming dynamic power as no active circuitry is used. In the interest of minimizing power, and avoiding the design of a power on/off mechanism for an active CMFB circuit, a passive CMFB approach has been taken for the opamps. The passive CMFB approach used is shown in Fig. 7.32, and exploits charge conservation to set the output common mode to $V_{DD}/2$ [37].

For the circuit of Fig. 7.32, during φ_1(power-off phase: voutp = voutn = V_{DD}), assuming $C_1 = C_2 = C_3$:

$$Q_{C1} = Q_{C2} = C(V_{DD} - V_B) \tag{7.8}$$

$$Q_{C3} = -CV_B \tag{7.9}$$

and during $\overline{\phi_1}$:

$$Q_{C1} = C(V_{outp} - V_{CM}) \tag{7.10}$$

$$Q_{C2} = C(V_{outn} - V_{CM}) \tag{7.11}$$

$$Q_{C3} = C(V_{DD} - V_{CM}) \tag{7.12}$$

Θ Charge is conserved: $V_{DD} - 3V_B = (V_{outp} + V_{outn}) - 3V_{CM}$

If V_B is set to approximately the intended value of V_{CM} (determined through desired bias current): $\Rightarrow \frac{(V_{outp}+V_{outn})}{2} = V_{output=CM} = \frac{V_{DD}}{2}$

7.10 Power Reduction Through Current Modulation

As described in Section 7.8, the Rapid Power-On Opamp developed for the power scalable ADC of this chapter has the advantage of short power on/off times due to a bias isolating-feedback based biasing architecture. The Rapid Power-On Opamp can also be used in other applications where analog blocks requiring opamps do not always need to be powered on. For example in a conventional pipelined ADC using an MDAC architecture similar to Fig. 3.5, the opamp is only required during the second clock phase (hold phase). If the opamp is powered-off during the first clock phase (sampling phase), the average MDAC power can be reduced by the fraction of time it is off over the total time the MDAC is operative. Ideally powering off the opamp for half a clock cycle affords a reduction in opamp power (which is a large portion of the entire ADC power) by 50%, thus allowing for substantial reductions in analog power consumption as shown in Fig. 7.33.

As Rapid Power-On Opamps allow for reduced power consumption in pipeline ADCs, the MDACs of the pipelined ADC in this chapter have been designed to only

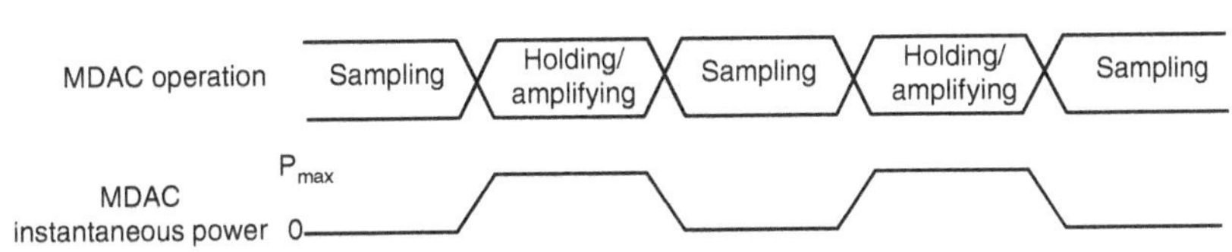

Fig. 7.33 Illustration of MDAC Power reduction using rapid power-on opamp

power-on during the hold phase, when using the hybrid CMPS power scale technique, thereby reducing power consumption. To evaluate the power reduction afforded by powering off the MDACs during the sampling phase, an additional mode of operation has been designed, where the opamps are never powered-off, i.e. the ADC operates as a conventional pipelined ADC. Thus the ADC has three modes of operation:

Mode 1 (*power scalable mode*): Pipelined ADC uses CMPS to allow for power to be scaled with sampling frequencies for low sampling rates, but has a maximum sampling rate limited by the pipeline latency as described in Section 7.5. The MDAC opamps are only powered on during the hold phase.

Mode 2 (*power reduction mode* – PRM): Pipeline ADC uses pipeline architecture to operate at maximum speed. Current scaling is used to reduce power with sampling frequency over a narrow range of sampling rates not covered by CMPS as described in Section 7.5. To minimize power consumption the MDAC opamps are powered off during the sampling phase, exploiting the short power-on times of the Rapid Power-On Opamp.

Mode 3 (*nominal mode* – NM): Same as Mode 2, except the MDAC opamps are always powered on.

Thus by measuring the ADC power in power reduction mode and nominal mode, it is possible to determine the amount of power savings afforded by powering-off the MDAC opamps during the sampling phase. It is not possible to compare the power of nominal mode with power scalable mode as the power scalable mode explicitly requires opamps to power-off to achieve power scalability.

The exact mode of ADC operation was set via off chip control bits. The MDAC opamps in power reduction and power scalable modes were powered on at the same time the advanced sampling clock of the MDAC bottom plate samples, to ensure the opamp is powered on before a virtual ground is required. Thus the opamps are on for larger than 50% of the period (i.e. $T/2 + t_{non\text{-}overlap}$). For higher sampling rates (>50 MS/s) where $t_{non\text{-}overlap}$ (shown in Section 7.15 to be 1.4 ns) is comparable to T/2, the effect of power reduction is less pronounced, as the opamps are powered-off for a shorter portion of the clock period. A more aggressive design can significantly reduce $t_{non\text{-}overlap}$, thus allowing for near 50% reduction in opamp power for higher sampling rates.

It should be noted that in several high speed analog applications there exist idle time slots where opamps are not required (e.g. discrete time filters, sample-and-holds, etc.). By using Rapid Power-On Opamps in such applications it is possible to significantly reduce the analog power by completely powering off the opamps when not required.

7.10.1 Common Mode Feed Back (CMFB) for Different Opamp Modes

As described in Section 7.10, the ADC operates in three modes. As such the CMFB must facilitate the various constraints in each mode: in power scalable and power

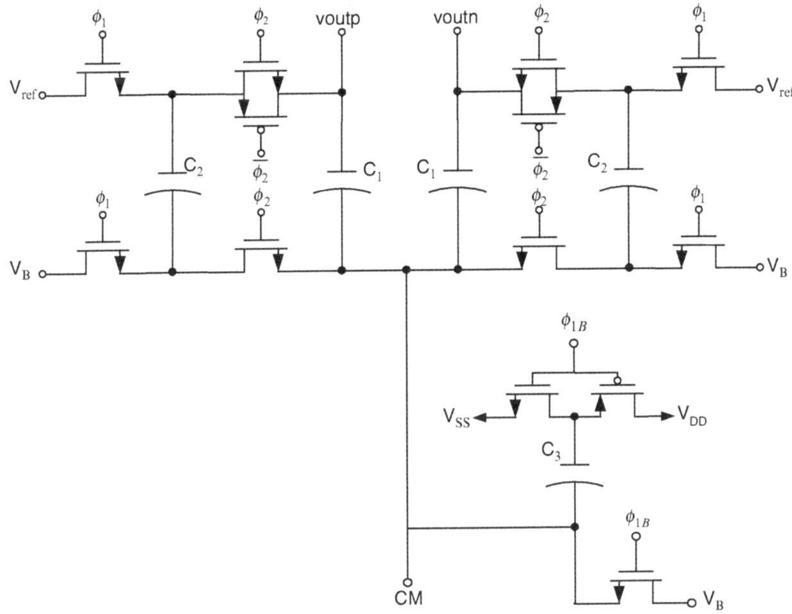

Fig. 7.34 Hybrid switched capacitor CMFB circuit

reduction modes, the CMFB of Fig. 7.32 can be used, whereas in nominal mode, the opamps are always on, thus do not have a reset phase hence cannot use the CMFB of Fig. 7.32. For nominal mode operation, a conventional switched capacitor CMFB structure of Fig. 7.31 is required. To allow for a single CMFB circuit to work for all desired operating modes, a reconfigurable CMFB circuit was used which changes its structure depending on the mode of ADC operation as shown in Fig. 7.34.

7.11 Sample-and-Hold (S/H)

From the discussion of Section 3.10, a front-end S/H is required to ensure the sub-ADC and MDAC of the first stage see the same input for high input frequencies. In this work a front-end S/H topology as shown in Fig. 7.35 was used [25].

As the S/H is the first stage in the pipeline, the S/H opamp bandwidth must be at least that of the first stage. To minimize the design/layout time, the opamp used in stage one is reused in the S/H. A transmission gate was used for S1, where the MOS switch sizes were sized such that the S/H on it is own had an input referred SFDR > 72 dB for input frequencies greater than 50 MHz (NMOS W/L = 2 μ/0.18 μM = 30, PMOS W/L = 4 μ/0.18 μM = 30). Simulations showed sufficient SFDR using transmission gates for the input switch S1, thus more complicated boot-strapped techniques were not required.

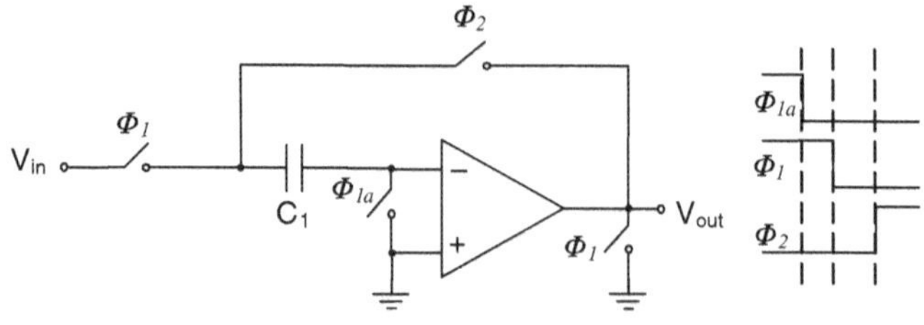

Fig. 7.35 Front-end S/H (shown single-ended, but implemented differentially)

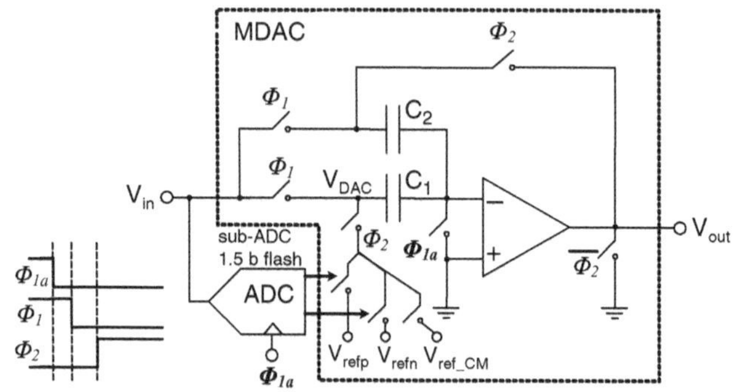

Fig. 7.36 1.5-bit Pipelined stage architecture (shown single-ended, implemented differentially)

7.12 1.5-bit MDAC

The 'flip-around' MDAC as shown in the following figure was used for the pipeline ADC due to the large feedback factor, hence fast transient response, and good matching (due to the identical capacitor sizes, i.e. $C_1 = C_2$). The sizes used for the sampling and feedback capacitors are listed in Table 7.2. To maximize capacitor matching, and minimize absolute variation, MIM capacitors were used. Switches for the MDAC were sized to meet the minimum RC time constant for the maximum sampling frequency (>50 MS/s), where transmission gates were used if the signal to be passed included $V_{DD}/2$ (Fig. 7.36).

7.13 Sub-ADC Comparators

From Section 3.8, since digital error correction allows for less accurate comparators in the sub-ADC, dynamic comparators were used to implement the 1.5-bit flash ADC in each pipelined stage. Dynamic comparators have the advantage of low power consumption, but at the cost of increased offset. This however is a favorable tradeoff as the 1.5-bit topology has a large amount of redundancy to trade with

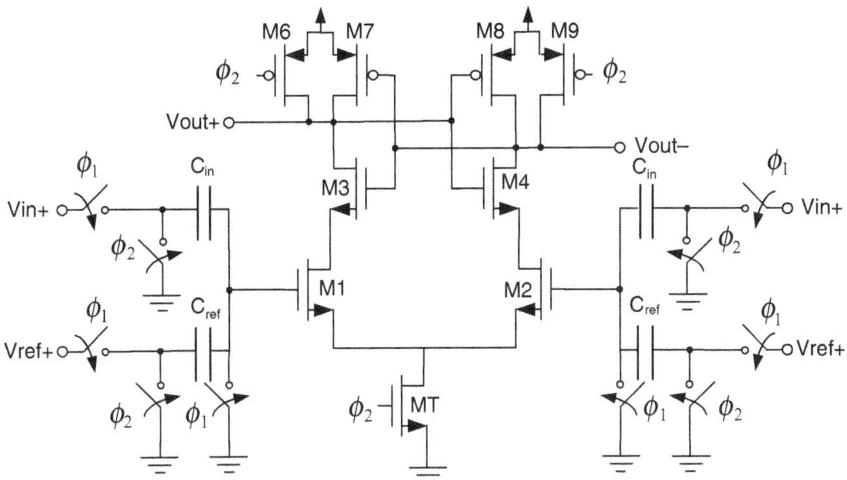

Fig. 7.37 Dynamic comparator used in flash sub-ADC

comparator offset [15]. The following figure illustrates the comparators used in this work (Fig. 7.37).

7.14 Bias Circuits

From Fig. 7.24 opamp bias circuits are required to provide a cascode bias to each opamp. Bias voltages for cascode opamps are typically derived from wide-swing cascode current mirrors [10]. As mentioned in Section 7.5 however, a small amount of current scaling is used in this design so as to provide a continuous power scalable range. The ADC should also be able to be biased deep in weak inversion so that the problems of increased bias-point sensitivity and mismatch can be empirically measured. As such the bias circuit must keep M2 and M3 in Fig. 7.38 in the active region regardless of the bias current (and thus level of channel inversion).

In [32] a wide swing cascode bias circuit (Fig. 7.38) was shown to ensure the opamp cascode transistors remain active so long as one of M2 and M3 are in strong inversion. If both transistors fall into weak inversion however (which is inevitable for a sufficient current scaling), the required ratio of W_{M1}/W_{M2} to maintain both M2 and M3 in saturation would be excessive [32]. Subsequently rather than using a diode-connected transistor to bias M2, an alternative biasing network must be used to maintain active operation over wide variations in current. An alternative architecture that provides active cascode biasing that is independent of bias current is presented in [40], and illustrated in Fig. 7.39.

A detailed functional analysis of the bias circuit of Fig. 7.39 can be found in [40], however it is noted here the level of device saturation is set by the ratio of device widths of M2 and M3. The architecture is such that M4 and M5 stay in saturation

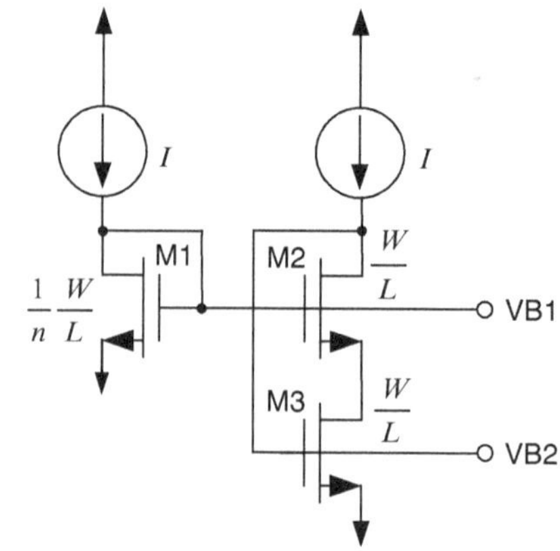

Fig. 7.38 Wide swing cascode current mirror (n is typically > 4)

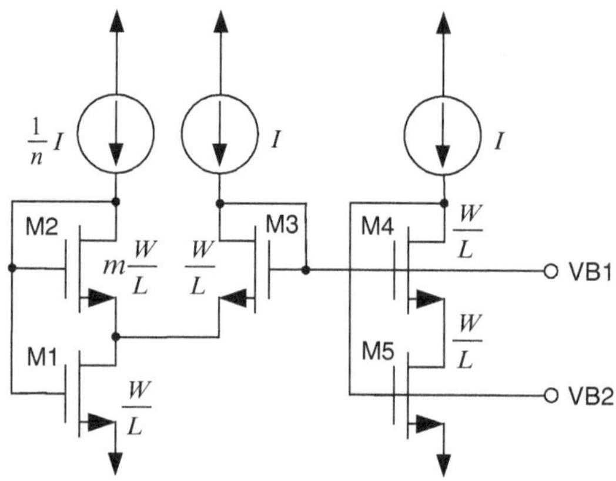

Fig. 7.39 Inversion insensitive bias circuit

regardless of channel inversion (i.e. weak or strong inversion). An off chip resistor was used to provide a constant current reference, where the resistor biased an on-chip PMOS current mirror. As described in Section 7.6, the power to the bias circuits was modulated using a series current switch approach.

7.15 Non-overlapping Clock Generator

Non-overlapping clocks are required in the MDAC to minimize the effects of signal-dependent charge injection. Non-overlapping clocks were generated using the design of Fig. 7.40 [42], where the non-overlap time is given by the minimum delay of t_2 and $t_2 + t_3 - t_5$ [42].

For this design, to improve the likelihood of design functionality (i.e. ensure enough time is given for clocks to fully swing rail to rail, and comparators to latch),

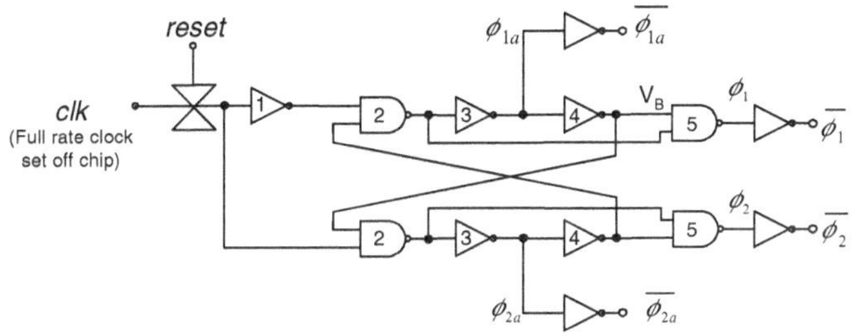

Fig. 7.40 Non-overlapping clock generator

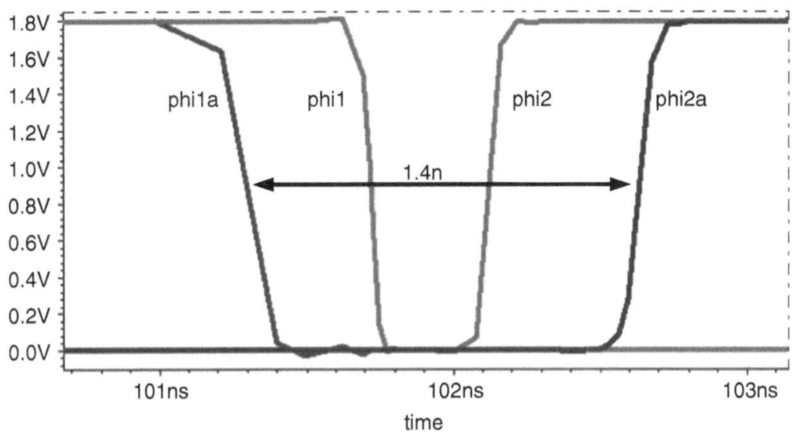

Fig. 7.41 Illustration on non-overlapping time in SPICE simulation

a long non-overlap time has been favored (shown in Fig. 7.41 to be 1.4 ns). For higher sampling rates, where 1.4 ns comprises a significant percentage of the settling time (e.g. for 50 MHz, half pulse width is only 10 ns), the settling time is reduced. In commercial designs, the non-overlap time can be carefully optimized to maximize the settling accuracy, such that a minimum power is required achieve the desired settled accuracy.

The non-overlapping clock generator was powered on/off via transmission gate at the clock input, which is enabled or disabled according to Fig. 7.16 by the signal 'reset'

7.16 Reference voltages

A significant advantage of the pipeline ADC architecture is the minimal use of reference voltages. Only three reference voltages are required for the entire ADC: a differential reference for the stage ADCs ($V_{refp} = 1.3$ V, $V_{refn} = 0.5$ V), and a common mode reference ($V_{ref\_cm} = 0.9$ V) for the CMFB circuit. In commercial designs the reference voltages are typically generated on chip through band gap circuits, and/or resistor ladders. To enhance testability and minimize on chip complexity however, the three reference voltages are generated off chip. Thus the reference voltages are not controlled by the state machine, hence in the implementation of this chapter, are not power scalable. Future work could investigate power on/off schemes for the reference voltages using the various techniques described in this chapter thus far.

7.17 Digital Error Correction

As the ADC of this chapter is a prototype, testability takes precedence over form. Thus rather than performing an on chip digital error correction, the outputs of each stage have been routed off-chip where error correction can be performed through a software post-processor (i.e. capture each stage digital output, and process it in Matlab to obtain the corrected 10-bit output). With the output of each stage available off chip, the design lends itself to more testability; if any errors are present in the design it is easier to debug where along the pipeline the problems are. As digital error correction typically consumes less than 5–10% of the total power budget, the exclusion of the block is not significant.

7.18 Experimental Implementation: PCB

A four layer FR4 dielectric PCB board with a minimum 6 mil trace was designed as shown in the following figure for the device under test. Separate Power planes were used to isolate the analog, digital, I/O, and board power supplies. A differential input was generated using a 1:1 turns ratio Minicircuits transformer matched to 50 Ω. Reference voltages ($V_{refp} = 1.3$ V, $V_{refn} = 0.5$ V, and $V_{ref\text{-}cm} = 0.9$ V) were generated by passing the output of a resistive voltage divider through an opamp (LM7301) in a unity gain buffer configuration. To maintain constant supply voltages, all voltage supplies for each power plane were generated through regulators (LM337, LM1117), and heavily decoupled with capacitors. As the ADC utilized a constant current biasing scheme, an off chip adjustable resistor was used as the master current source. The resistance was a series combination of 1, 10, 200 kΩ, 1 and 3 MΩ potentiometers such that the biasing current could be accurately controlled over a wide range to facilitate the evaluation of wide range current scaling (Fig. 7.42).

7.19 Experimental Implementation: Test Setup

A test setup as shown in Fig. 7.43 was used. Sinusoidal inputs were generated using a Rohde & Schwarz SMT03 function generator. A Minicircuits low pass filter was

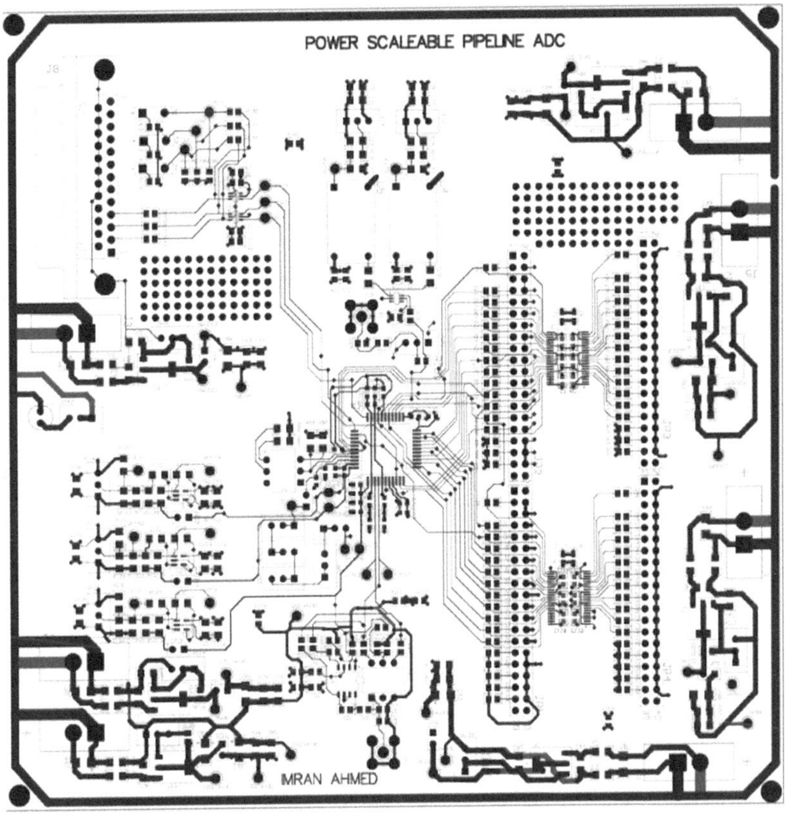

Fig. 7.42 Custom PCB layout

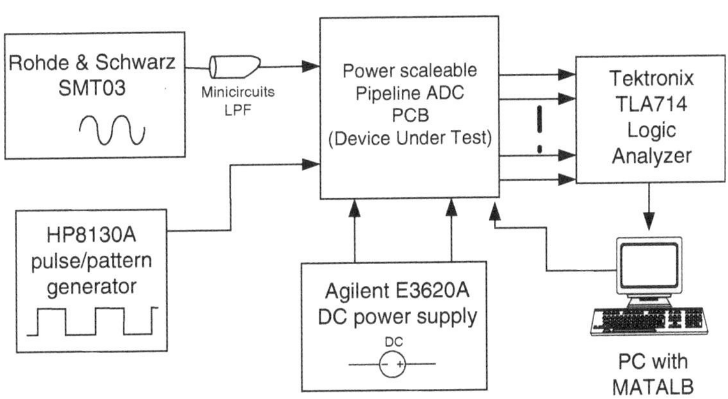

Fig. 7.43 Test setup for power scalable pipeline ADC

used to minimize harmonic distortion from the function generator such that the sinusoidal input to the ADC had an SNDR of well over 62 dB. An HP 81120A pulse/pattern generator was used to generate the clock to the ADC. The serial shift register was loaded via a parallel port connection to a PC, where a Matlab script was executed to load the appropriate bits. The output bits of each pipeline stage were captured using a Tektronix TLA714 logic analyzer, capable of capturing 65,536 points at a time. An Agilent E3620A Dual output DC power supply was used to provide positive and negative voltages to the voltage regulators on the PCB. The 10-bit output word from the 10-bit ADC was determined via a Matlab script written to emulate the operation of a digital error correction circuit.

7.20 Measured Results

The power scalable pipeline ADC was implemented in a 1.8 V, 0.18 μm CMOS process with single poly, six-metal layers, MIM capacitors, and Deep N-Well layer options. The core area was 1.1 × 1.1 mm (1.21 mm^2), and the total area including I/O drivers and bonding pads was 1.5 × 1.5 mm (2.25 mm^2). The integrated circuit was packaged in a 44-pin CQFP package. To minimize power supply related noise, analog pins were separated from digital pins on the power supply ring surrounding the ADC core. The layout of the fabricated chip is shown in Fig. 7.44, where key circuit blocks of the pipeline ADC have been highlighted.

As described in Section 7.5, the fabricated ADC achieves power scalability by using a hybrid CMPS technique, where current scaling is used to achieve power scalability for sampling rates not allowable with CMPS applied to a pipeline ADC. Since the CMPS technique provides power scalability by effectively multiplying the power scalable range achievable through current scaling while preserving the accuracy (as bias currents are unchanged when CMPS is enabled), the complete performance of the ADC can be quickly characterized by measuring the accuracy of the ADC for a narrow range of current scaled sampling rates. As such the measured results of the fabricated ADC are presented in two Sections (7.21 and 7.22). Section 7.21 presents measured results of the ADC for a small range of sampling rates, where current scaling is used to achieve power scalability. To evaluate the benefits of powering off the MDAC during the sampling phase, the power of the ADC in power reduction and nominal modes is also compared. The problems of extended power scaling using frequency dependent biasing are also elaborated with measured results in Section 7.21.2. Section 7.22 presents measured results showing the achievable power scalable range with CMPS applied to the current scaled sampling rates presented in Section 7.21, thereby validating that the accuracy is indeed preserved when CMPS is used for lower sampling rates, and that an extended power scalable range of sampling rates can be achieved without further reductions in bias currents.

Fig. 7.44 Photograph of fabricated chip

7.21 Current Scaled Power

The bias currents were scaled for f_s between 1 and 80 MS/s, where the current was set such that the figure-of-merit (see Section 2.4) was optimal. For each f_s, a full scale (1.6 Vpp) sinusoid near Nyquist was applied to the ADC input, where the SNDR, SFDR, and associated power were measured for each f_s.

To measure the benefit of the power reduction mode, for each current scaled sampling rate the ADC was measured in both PRM and NRM operation modes, where between each mode the bias currents were kept constant. The measured results of the ADC in PRM and NM are shown in Table 7.3, where Figs. 7.45–7.47 graphically illustrate the differences in accuracy and power of the two operation modes. The 65,536 point FFTs of the digitized output are shown in Figs. 7.48–7.53 for sampling rates of 50, 30 and 10 MS/s. Table 7.4 tabulates the figure of merit of the ADC for different current scaled sampling rates.

Table 7.3 Measured ENOB and power from fabricated ADC

Accuracy and power measurement of ADC									
f_s	f_{in}	SNDR (dB)		ENOB (bits)		SFDR (dB)		Power (mW)*	
(MS/s)	(MHz)	PRM	NM	PRM	NM	PRM	NM	PRM	NM
1	0.17	54.9	55.3	8.8	8.9	63	65	0.7	1.1
10	4.75	56.4	55.3	9.1	8.9	71	64	5.6	7.2
20	9.54	55.9	54.7	9.0	8.8	70	62	11.7	15.6
30	14.01	55.7	54.1	9.0	8.7	71	60	19.1	25.1
40	19.01	54.7	53.3	8.8	8.6	67	60	26.5	34.4
50	20.94	54.8	51.8	8.8	8.3	67	58	34.9	44.2
60	20.94	52.2	48ss.8	8.4	7.8	62	54	43.1	53.8
70	20.94	46.7	44.2	7.5	7.0	56	49	45.0	53.6
80	20.94	42.7	42.1	6.8	6.7	52	51	54.7	64.8

(PRM = Power reduction mode, NM = Nominal mode)
*Note power listed excludes I/O buffer, reference voltage, and digital error correction power

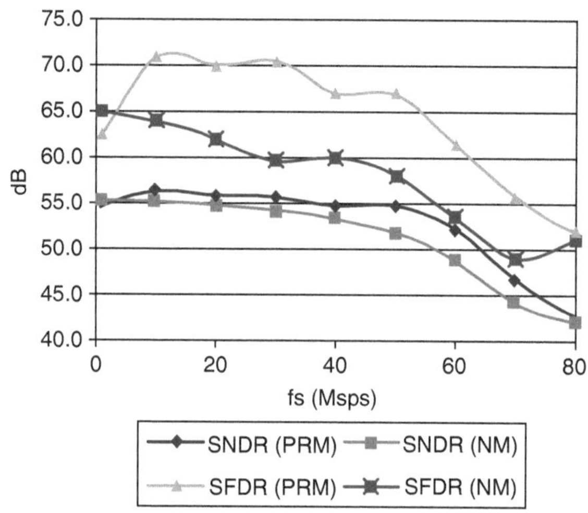

Fig. 7.45 SNDR, SFDR variation with sampling rate for PRM and NM

When the power reduction mode is enabled the power was reduced by as much as 30%. As described in Section 7.10, the relative reduction of power in the power reduction mode is less for higher sampling rates as the opamp is off for a smaller percentage of the clock cycle. As the ADC maintains similar accuracy between power reduction and nominal modes for f_s up to 80 MS/s, it can be concluded the effect of increased power supply noise due to ground bounce is negligible. The relatively low power supply noise is likely due to the use of large on chip supply-decoupling capacitors and large off chip capacitors to maintain a constant power supply. The power reduction method shows a slightly higher ENOB for faster sampling rates as the CMFB circuit for the opamp in power reduction mode operates on an advanced clock (~1.4 ns advanced from the clocks supplied to the

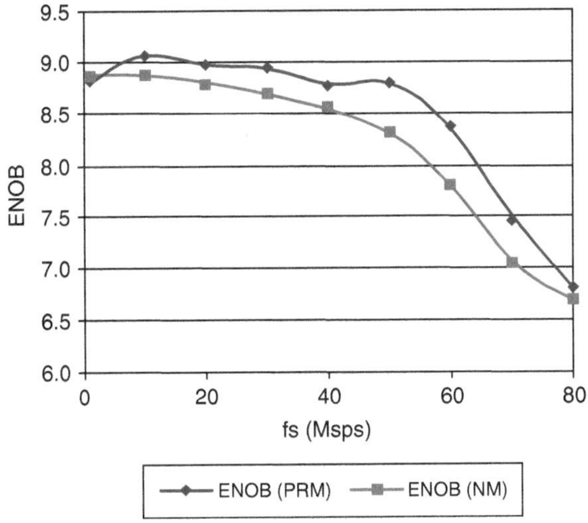

Fig. 7.46 ENOB variation with sampling rate for PRM and NM

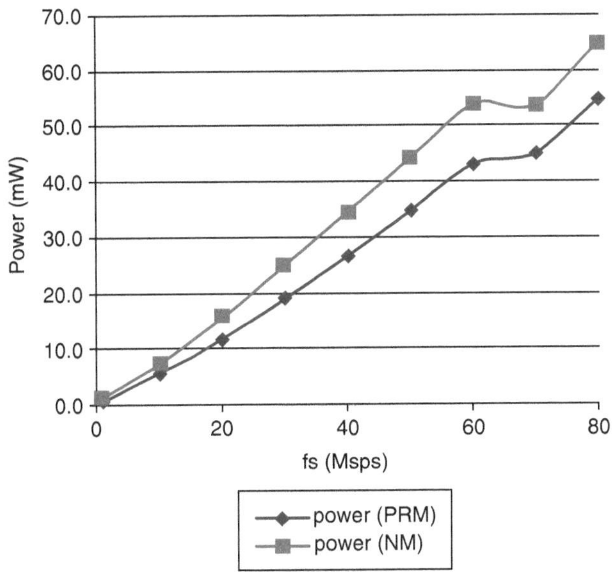

Fig. 7.47 Variation of power with sampling rate for PRM and NM

CMFB in nominal mode, thereby increasing the settling time), which for higher frequencies is a large portion of the settling time. If however the ADC in nominal mode is supplied additional current, measured results show the ADC settles to the same accuracy in both cases. The similar accuracy and significantly reduced power

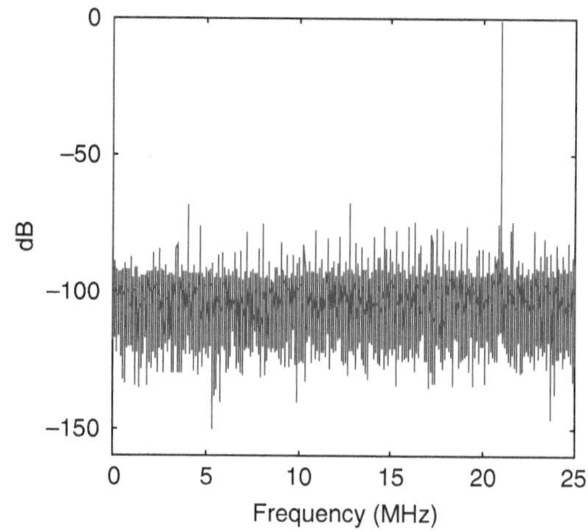

Fig. 7.48 $f_s = 50$ MS/s, $f_{in} = 20.9371$ MHz, PRM

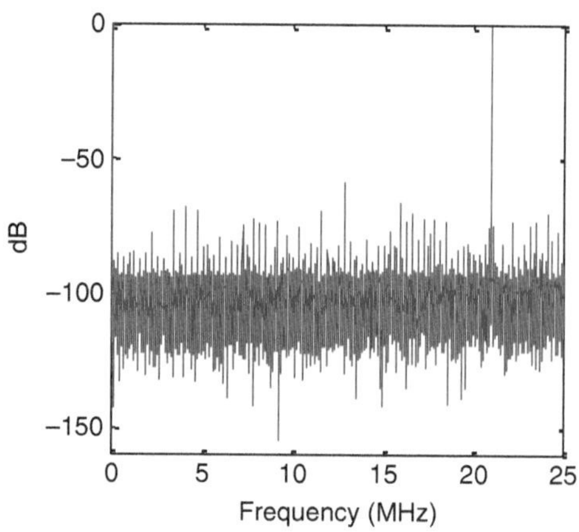

Fig. 7.49 $f_s = 50$ MS/s, $f_{in} = 20.9371$ MHz, NM

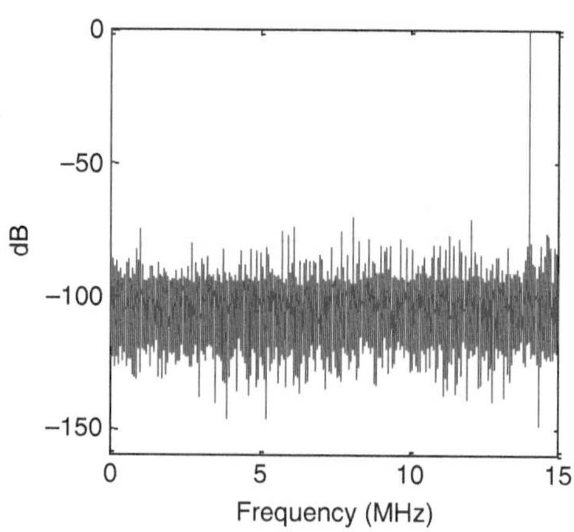

Fig. 7.50 $f_s = 30$ MS/s, $f_{in} = 14.013$ MHz, PRM

Fig. 7.51 $f_s = 30$ MS/s, $f_{in} = 14.013$ MHz, NM

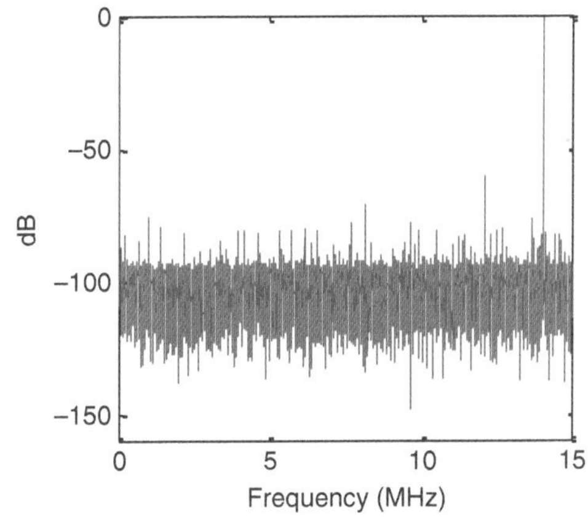

Fig. 7.52 $f_s = 10$ MS/s, $f_{in} = 4.571$ MHz, PRM

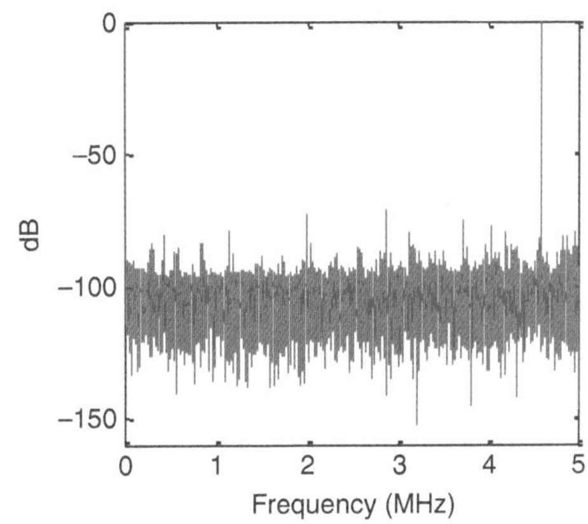

Fig. 7.53 $f_s = 10$ MS/s, $f_{in} = 4.571$ MHz, NM

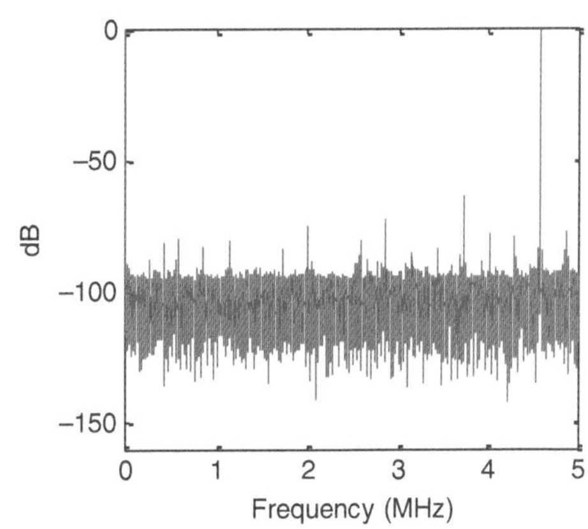

Table 7.4 Figure of merits for measured ADC at various f_s

Accuracy and power measurements of ADC						
f_s (MS/s)	f_{in} (MHz)	FOM (pJ/step) PRM	FOM (pJ/step) NM	FOM (mW/Msps) PRM	FOM (mW/Msps) NM	Power reduction (%)
1	0.17	1.6	2.2	0.72	1.06	32.3
10	4.75	1.0	1.5	0.56	0.72	22.0
20	9.54	1.2	1.7	0.59	0.78	24.9
30	14.01	1.3	2.0	0.64	0.84	23.9
40	19.01	1.5	2.3	0.66	0.86	23.0
50	20.94	1.6	2.8	0.70	0.88	21.1
60	20.94	2.2	4.0	0.72	0.90	19.9
70	20.94	3.6	5.8	0.64	0.77	16.1
80	20.94	6.1	7.8	0.68	0.81	15.6

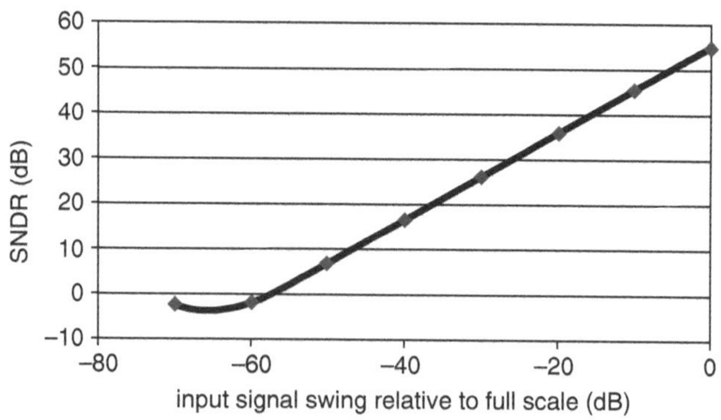

Fig. 7.54 Input dynamic range, $f_s = 50$ MS/s, $f_{in} = 20.371$ MHz

in power reduction mode demonstrates the effectiveness of using Rapid Power-On Opamps to significantly reduce power in switched capacitor circuits at high sampling rates.

The ADC is further characterized in Figs. 7.54–7.60 at current scaled sampling rates of 50, 30, and 10 MS/s. Only results from power reduction mode are shown as the power reduction mode produces the best figure of merit. The plots include the variation of SNDR with supply voltage, and SNDR with input signal swing. As the input bandwidth of the ADC is limited by the layout and input switch sizes only (since a S/H is used on the front end of the ADC), the variation of SNDR with input frequency is only presented for $f_s = 50$ MS/s.

Of note, when the supply voltage was reduced, the master bias current was re-adjusted for each V_{DD} to be constant (if possible) to maintain fixed opamp bandwidths, and the reference voltages adjusted to maintain a fixed total V_{DS} across the opamp transistors of 800 mV (i.e. input signal swing for $V_{DD} = 1.8$, 1.7, 1.6, 1.5, 1.4, 1.3 was V_{refp}–V_{refn} = 1.6, 1.6, 1.6, 1.4, 1.2 and 1 V peak-to-peak respectively). The input frequency was not adjusted between V_{DD} measurements for fixed f_s.

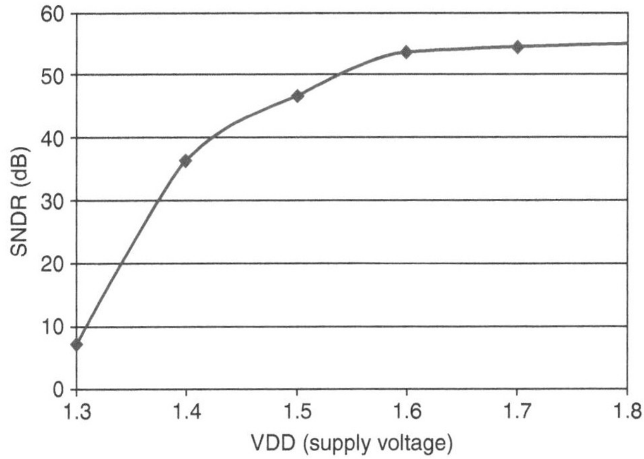

Fig. 7.55 SNDR versus supply voltage for $f_s = 50$ MS/s, $f_{in} = 20.173$ MHz

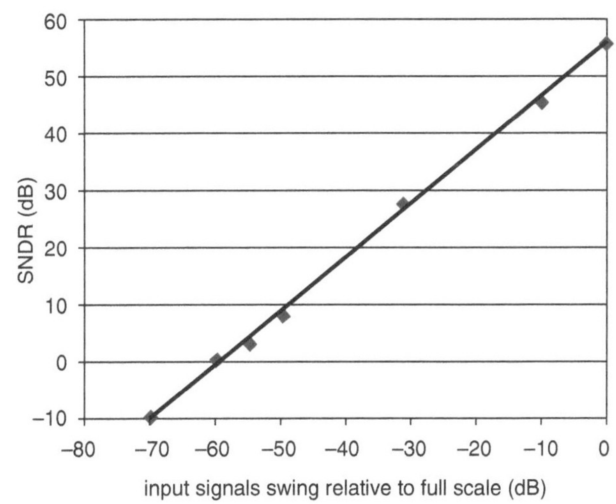

Fig. 7.56 Input dynamic range, $f_s = 30$ MS/s, $f_{in} = 14.317$ MHz

7.21.1 Power Reduction Mode: Static Accuracy

A histogram-based approach was taken to measure the INL and DNL of the ADC [43]. The histogram approach has the advantage of measuring ADC linearity without resorting to more time consuming DC approaches, which require several thousand different measurements, necessitating a more complicated automated testing routine. The histogram approach however has the disadvantage of requiring a large number of samples to measure the INL/DNL to at least 0.1 LSB accuracy

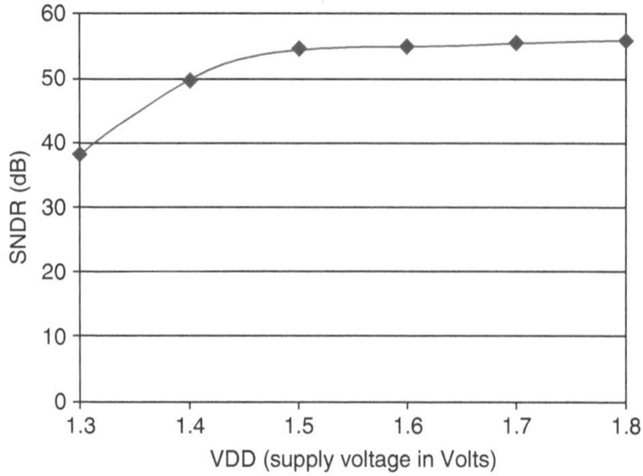

Fig. 7.57 SNDR versus supply voltage for $f_s = 30$ MS/s, $f_{in} = 20.173$ MHz

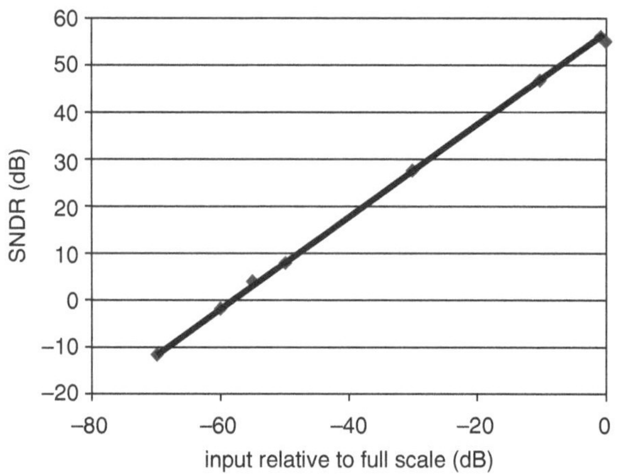

Fig. 7.58 Input dynamic range, $f_s = 10$ MS/s, $f_{in} = 4.571$ MHz

(>1,000,000 for 10-bit ADCs). As the Tektronix TLA714 logic analyzer only had a memory of 65,536 samples, to obtain an accuracy of at least 0.1 LSB in the INL/DNL plots, several measurement captures of the logic analyzer were combined together to form a single measurement file, which had 1,048,576 output samples. A script written in MATLAB was used to derive the INL and DNL from the amalgamated data capture. The INL and DNL plots of the ADC in power reduction mode at sampling rates of 10, 30 and 50 MS/s are shown in Figs. 7.61–7.66, where the maxima/minima are listed in Table 7.5.

As mentioned in Section 7.21 the power of the ADC was adjusted to achieve an optimal figure of merit. By providing a minimal power to the MDACs, some stages

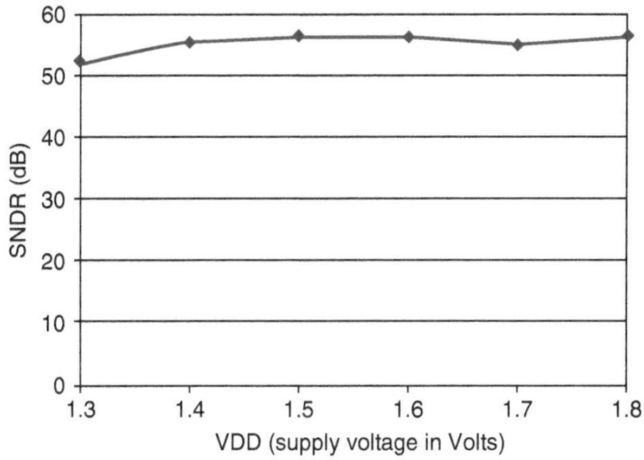

Fig. 7.59 SNDR versus supply voltage for, $f_s = 10$ MS/s, $f_{in} = 4.571$ MHz

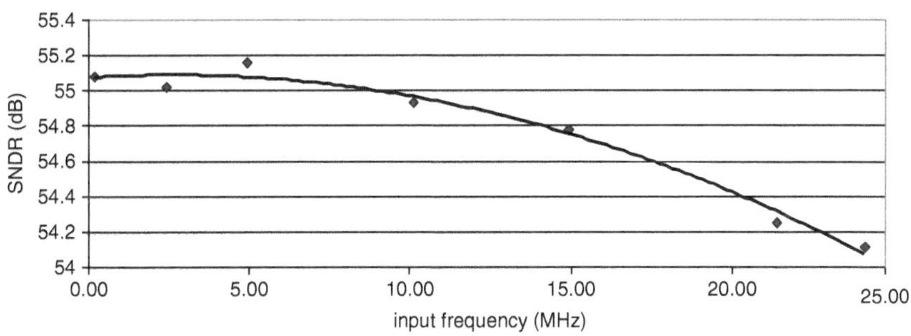

Fig. 7.60 SNDR versus. input frequency for $f_s = 50$ MS/s

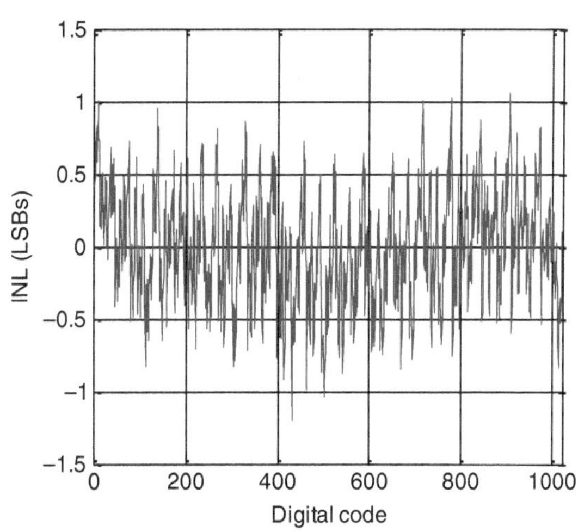

Fig. 7.61 INL @ 50 MS/s

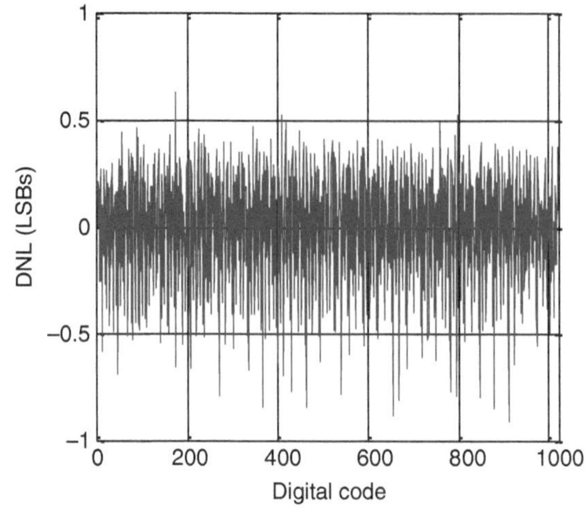

Fig. 7.62 DNL @ 50 MS/s

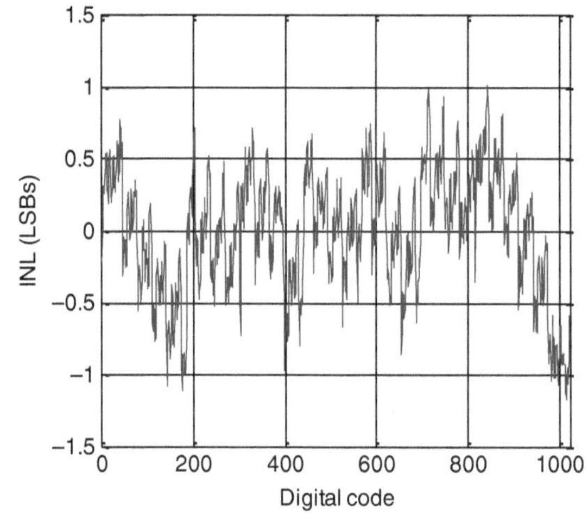

Fig. 7.63 INL @ 30 MS/s

can become bandwidth limited to less than 10-bit accuracy, hence inducing gain error due to incomplete settling. If the ADC power is increased well beyond the optimal power, such that the stages are not bandwidth limited to less than 10-bit accuracy, it can be evaluated if in fact the larger linearity errors are due to insufficient opamp bandwidth, or capacitor mismatch/insufficient DC opamp gain. The INL/DNL plots of the ADC with maximum power are shown in Figs. 7.67–7.72, where the maxima/minima are listed in Table 7.6:

As the ADC shows smaller linearity errors when not bandwidth limited, it can be inferred the larger linearity errors in Figs. 7.61–7.66 are due to some stages in the pipeline ADC being bandwidth limited when optimizing for figure-of-merit.

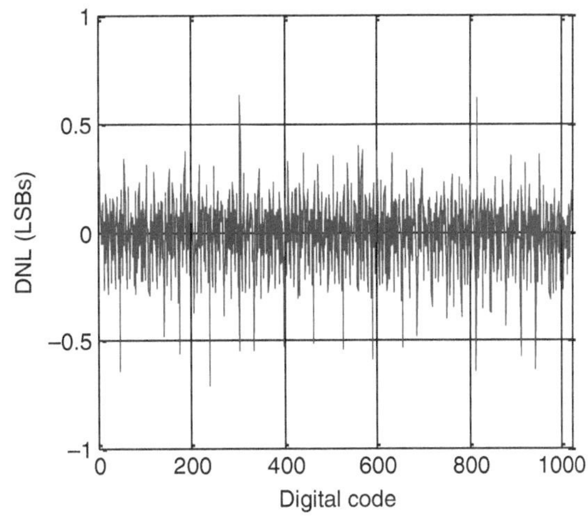

Fig. 7.64 DNL: @ 30 MS/s

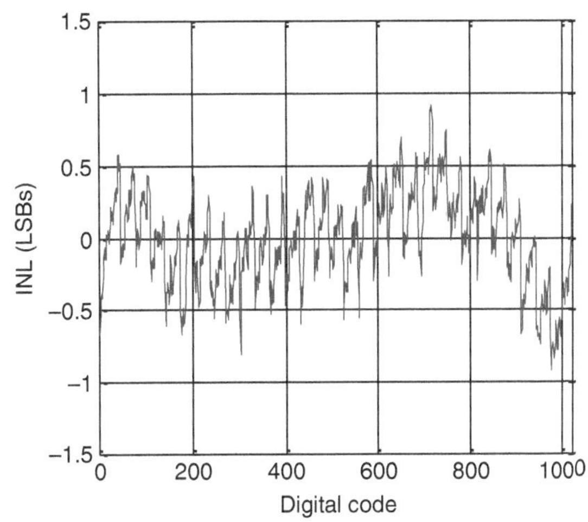

Fig. 7.65 INL @ 10 MS/s

7.21.2 Power Scalable ADC: Current Scaling

As described in Chapter 4, there are several problems in using current scaling as a technique to scale power with sampling frequency over a wide variation in sampling rates. The problems fall into two categories: pre-design and post-Design. Pre-design problems include increased design time due to multiple bias points, and more complicated and less reliable simulation models. Post-design problems include poorer yield, and increased bias point sensitivity. Only post-design problems may be empirically evaluated. For a proper yield analysis, the performance of the ADC biased deep in the weak inversion region must be measured for several

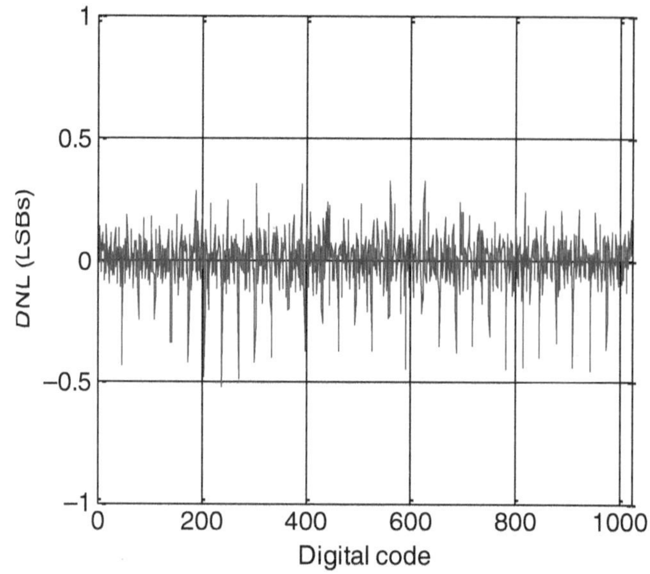

Fig. 7.66 DNL @ 10 MS/s

Table 7.5 INL/DNL maxima and minima for f_s = 10, 30, 50 MS/s for current scaled f_s

f_s (MS/s)	Max INL	Min INL	INL pp	Max DNL	Min DNL	DNL pp
10	+0.92	−0.92	1.84	+0.32	−0.52	0.84
30	+1.01	−1.17	2.18	+0.63	−0.70	1.33
50	+1.06	−1.19	2.25	+0.63	−0.91	1.54

hundred fabricated chips with identical setups (i.e. same input/clock frequency and supplied bias current). Such an analysis was not possible with the time, equipment, and number of chips fabricated available (only five packaged ICs were delivered) – thus a true yield analysis cannot be performed. Larger performance variations in weak inversion however, can be inferred from a bias point analysis of the ADC biased in weak inversion.

As described in Section 7.14, a constant current biasing scheme set by an off chip resistor was used for the ADC. By using a digitally controlled voltage source (Agilent E3631A) rather than an off chip resistor, the V_{GS} of the on-chip current mirror shown in Fig. 7.73 can be precisely controlled. As such the impact of small variations of the bias voltage V_B (i.e. V_{GS} of M2 in Fig. 7.73) on ADC accuracy can be measured.

The theory of Section 4.7 predicts the exponential dependency of drain-source current on gate-source voltages of devices in weak inversion cause the ADC to be more susceptible to bias fluctuations (i.e.) a larger reduction in performance for the same increment in bias voltage when the ADC is deeper into the weak inversion

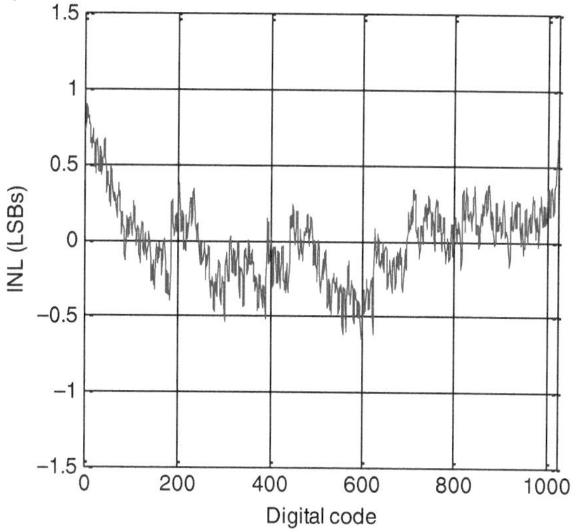

g. 7.69 INL @ 30 MS/s (Max BW)

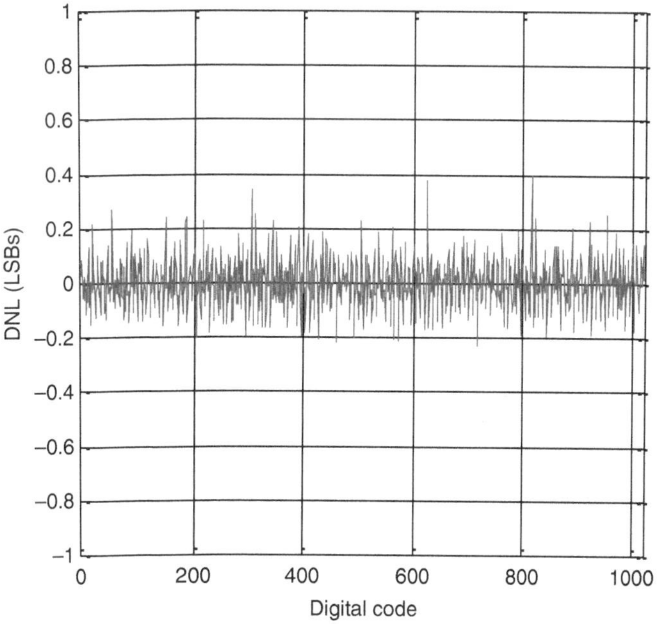

g. 7.70 DNL @ 30 MS/s (Max BW)

gion. To verify the claim, the ADC was biased with the digitally controlled oltage source such that an ENOB of 7–8 bits was achieved at f_s = 100 kS/s, 1,), 30 and 50 MS/s. The ENOB and power of the ADC were measured as the V_{GS}

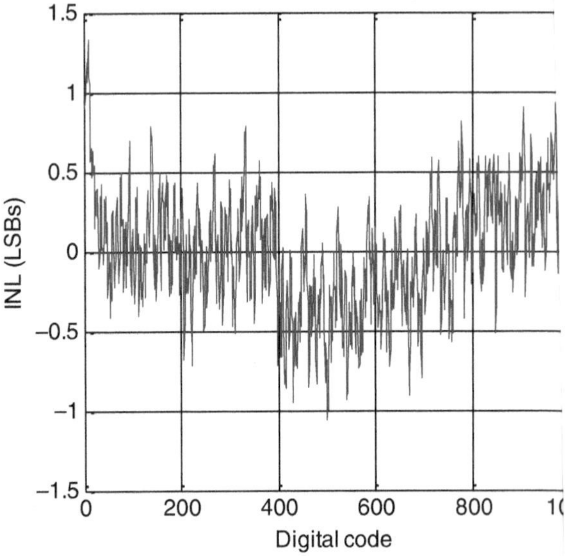

Fig. 7.67 INL @ 50 MS/s (Max BW)

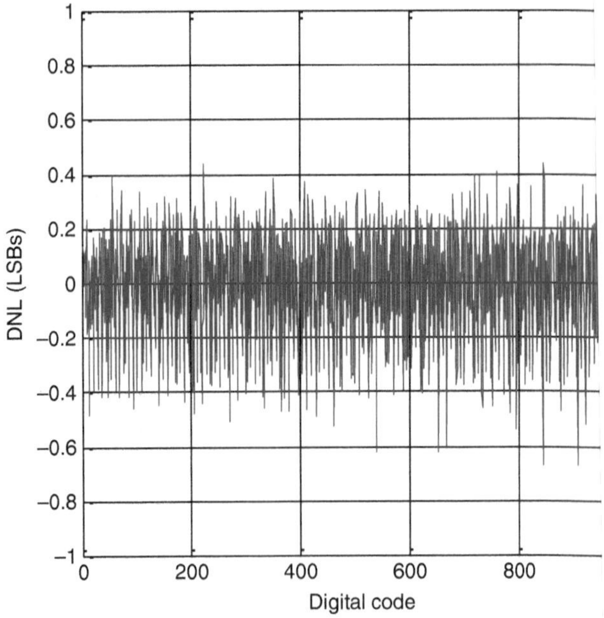

Fig. 7.68 DNL @ 50 MS/s (Max BW)

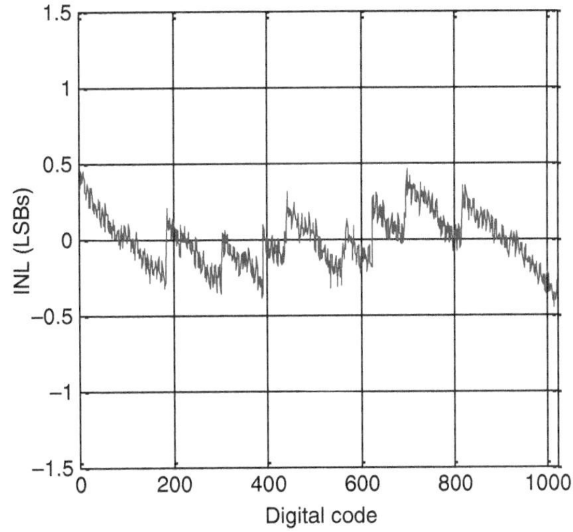

Fig. 7.71 INL @ 10 MS/s (Max BW)

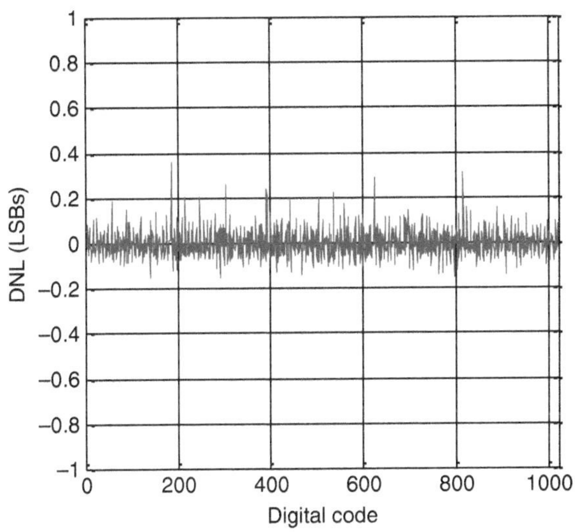

Fig. 7.72 DNL @ 10 MS/s (Max BW)

Table 7.6 INL/DNL maxima and minima for $f_s = 10, 30, 50$ MS/s for maximum bandwidth

f_s (MS/s)	Max INL	Min INL	INL pp	Max DNL	Min DNL	DNL pp
10	+0.46	−0.45	0.91	+0.36	−0.15	0.51
30	+0.90	−0.64	1.54	+0.40	−0.23	0.63
50	+1.34	−1.13	2.47	+0.44	−0.67	1.11

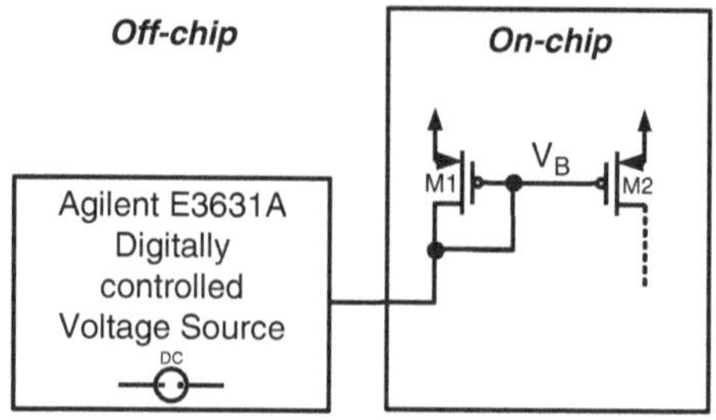

Fig. 7.73 Setup to perform bias point analysis

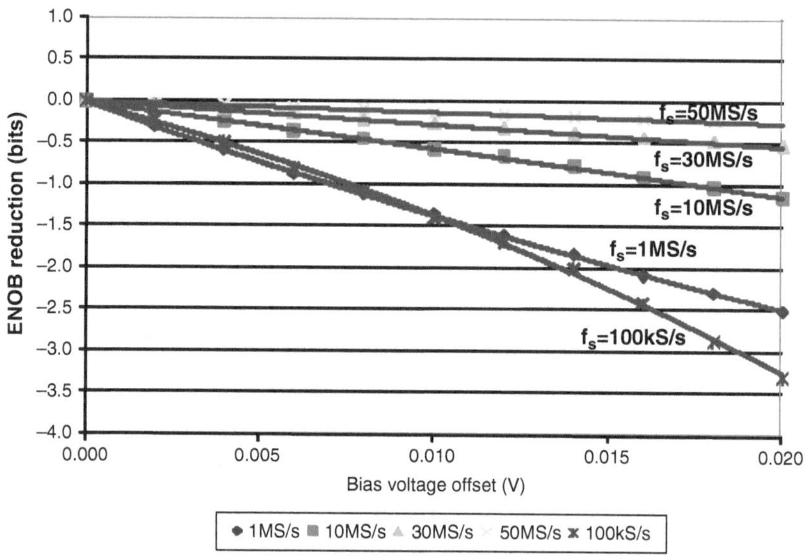

Fig. 7.74 Bias point sensitivity of ADC as current reduced with f_s

bias voltage of the on-chip current mirror was increased by 20 mV in 2 mV steps (thereby decreasing the on-chip V_{eff}). The measured ENOB reduction is shown graphically in Fig. 7.74.

Thus, although it is possible to obtain a high accuracy as the ADC is driven into the weak inversion region (e.g. can achieve >8 bits even for very low bias currents), the ADC performance becomes highly sensitive to bias voltage variations. It may also be inferred that any offset incurred due to a threshold mismatch would similarly affect ADC performance (e.g.) the reduction in performance due to a 5 mV threshold mismatch on the on-chip current mirror can be inferred from Fig. 7.74. Thus although it is possible to achieve a similar peak performance when current scaling is used to reduce analog power with sampling frequency, the increased bias sensitivity and poorer yield make current scaling over a large range impractical from a robustness standpoint.

7.22 Power Scalable ADC: Power Scaling Usingty CMPS

To evaluate the digitally controlled power scalable mode, clock frequencies between 1 and 50 MHz were supplied to the state machine, where the ADC power was scaled (by adjusting the bias currents) according to Table 7.3. The state machine was programmed to various effective sampling rates (i.e. between one seventh and 1/3,584th the clock supplied to the state machine), and the power measured for each effective sampling rate. The measured power, SNDR, and SFDR of the ADC using CMPS at clocks of 50, 30, 10 and 1 MHz supplied to the state machine (f_{sm}) are presented in Tables 7.7–7.10, and subsequently in graphical form (Figs. 7.75–7.82).

As predicted, the SNDR varied minimally over effective sampling rate, since the bias currents remain constant between sampling rates (only the off time changes, as described in Section 7.2). From the above figures, the ADC power is shown to scale with sampling frequency, where for lower sampling rates for a particular f_{sm}, the power scalable range was ultimately limited by the power of the digital state machine. As shown in Tables 7.7–7.10, the state machine power can be reduced by reducing the state machine clock speed, thereby lowering the lowest power scalable sampling rate. If the analog power is scaled with the clock supplied to the state machine so as to minimize analog power however, the lowest possible

Table 7.7 ADC performance using CMPS with $f_{sm} = 50$ MHz

f_s (Hz)	f_{in} (Hz)	P_{analog} (mW)	$P_{digital}$ (mW)	P_{Total} (mW)	SNDR (dB)
5.55E + 06	1.79E + 06	9.72	1.05	10.77	54.3
3.12E + 06	1.79E + 06	5.45	0.84	6.29	54.3
1.67E + 06	1.79E + 06	2.88	0.71	3.59	54.1
8.62E + 05	1.79E + 06	1.58	0.64	2.22	54.3
4.38E + 05	1.79E + 06	0.84	0.60	1.43	54.4
2.21E + 05	1.79E + 06	0.46	0.59	1.04	54.2
1.11E + 05	2.00E + 05	0.27	0.57	0.84	54.5
5.56E + 04	2.00E + 05	0.17	0.57	0.74	53.8
2.77E + 04	2.00E + 05	0.12	0.57	0.69	53.7
1.39E + 04	2.00E + 05	0.10	0.57	0.67	53.5

Table 7.8 ADC performance using CMPS with $f_{sm} = 30$ MHz

f_s (Hz)	f_{in} (Hz)	P_{analog} (mW)	$P_{digital}$ (mW)	P_{Total} (mW)	SNDR (dB)	SFDR (dB)
3.33E + 06	1.79E + 06	5.53	0.63	6.15	55.7	63
1.87E + 06	1.79E + 06	2.98	0.50	3.48	55.4	61
1.00E + 06	1.79E + 06	1.65	0.43	2.08	55.6	62
5.17E + 05	1.79E + 06	0.87	0.39	1.25	55.2	62
2.63E + 05	1.79E + 06	0.46	0.36	0.82	55.4	63
1.33E + 05	2.00E + 05	0.25	0.35	0.60	55.1	69
6.66E + 04	2.00E + 05	0.15	0.34	0.49	54.9	68
3.32E + 04	2.00E + 05	0.09	0.34	0.44	55.1	67
1.67E + 04	2.00E + 05	0.07	0.34	0.41	55.3	67
8.36E + 03	2.00E + 05	0.06	0.34	0.40	55.3	67

Table 7.9 ADC performance using CMPS with $f_{sm} = 10$ MHz

f_s (Hz)	f_{in} (Hz)	P_{analog} (mW)	$P_{digital}$ (mW)	P_{Total} (mW)	SNDR (dB)	SFDR (dB)
1.11E + 06	1.79E + 06	1.710	0.19	1.90	56.1	61
6.25E + 05	1.79E + 06	0.941	0.17	1.11	55.9	62
3.33E + 05	2.00E + 05	0.503	0.14	0.65	55.0	60
1.72E + 05	2.00E + 05	0.264	0.13	0.39	55.8	66
8.77E + 04	2.00E + 05	0.140	0.12	0.26	55.5	66
4.42E + 04	2.00E + 05	0.076	0.12	0.19	55.5	64
2.21E + 04	2.00E + 05	0.044	0.11	0.16	55.2	63
1.12E + 04	2.00E + 05	0.028	0.11	0.14	55.3	69
5.58E + 03	2.00E + 05	0.022	0.11	0.14	55.6	66
2.79E + 03	2.00E + 05	0.016	0.11	0.13	55.9	68

Table 7.10 ADC performance using CMPS with $f_{sm} = 1$ MHz

f_s (Hz)	f_{in} (Hz)	P_{analog} (mW)	$P_{digital}$ (mW)	P_{Total} (mW)	SNDR (dB)	SFDR (dB)
1.11E + 05	2.00E + 05	0.2086	0.021	0.230	55.5	64
6.25E + 04	2.00E + 05	0.1172	0.017	0.134	55.5	64
3.34E + 04	2.00E + 05	0.0628	0.015	0.077	55.6	64
1.72E + 04	2.00E + 05	0.0333	0.013	0.046	55.6	64
8.77E + 03	2.00E + 05	0.0176	0.012	0.030	55.5	64
4.42E + 03	2.00E + 05	0.0095	0.012	0.021	55.6	65
2.22E + 03	2.00E + 05	0.0054	0.012	0.017	55.4	64
1.12E + 03	2.00E + 05	0.0035	0.012	0.015	55.6	64
5.58E + 02	2.00E + 05	0.0041	0.012	0.016	55.5	64

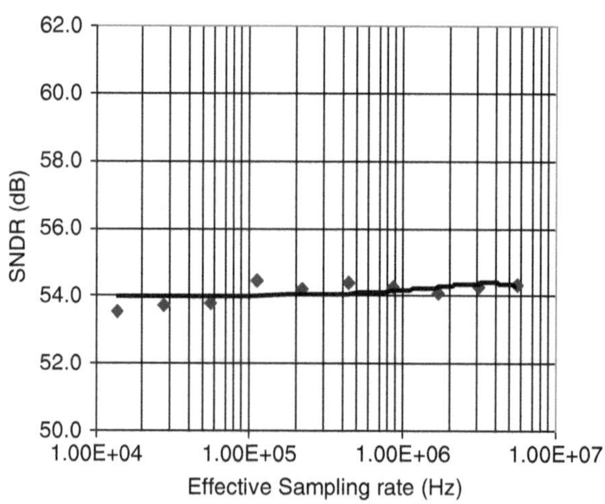

Fig. 7.75 SNDR variation with effective sampling rate for $f_{sm} = 50$ MHz

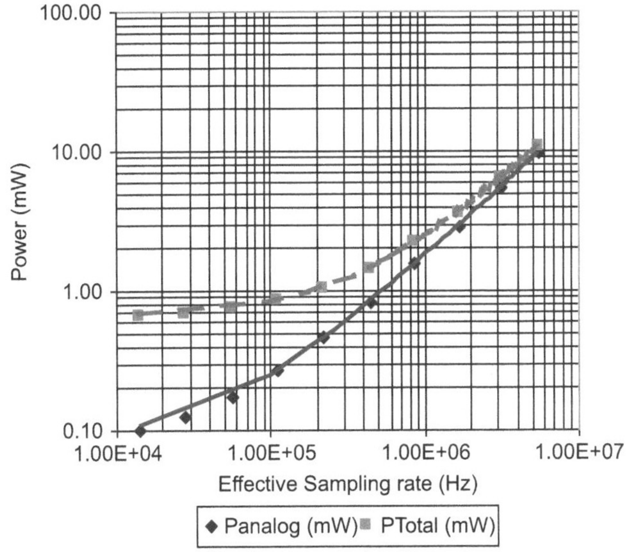

Fig. 7.76 Analog and total ADC power variation with effective sampling rate for $f_{sm} = 50$ MHz

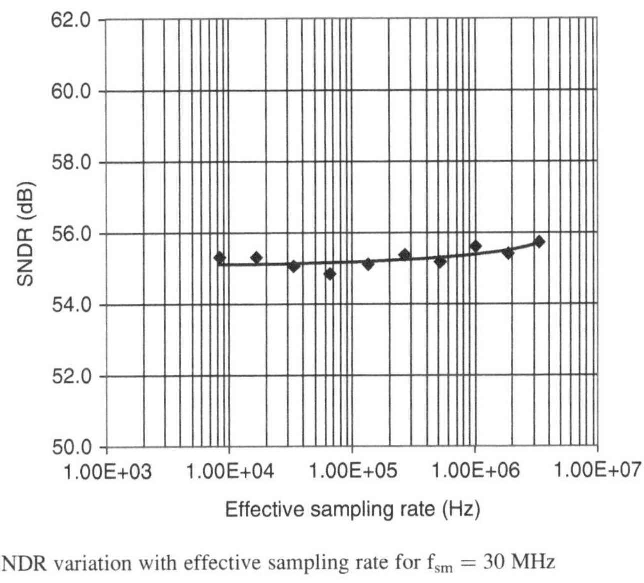

Fig. 7.77 SNDR variation with effective sampling rate for $f_{sm} = 30$ MHz

sampling rate achievable is limited by the sensitivity of the application in question to the problems of devices biased deep in weak inversion (as described in Chapter 4). (For example): If the yield, and bias point sensitivity of the ADC with bias currents adjusted to 1 MS/s were tolerable; the lowest sampling rate achievable from Table 7.10 is 580 Hz (~1:100,000 × power scalable range). However if the

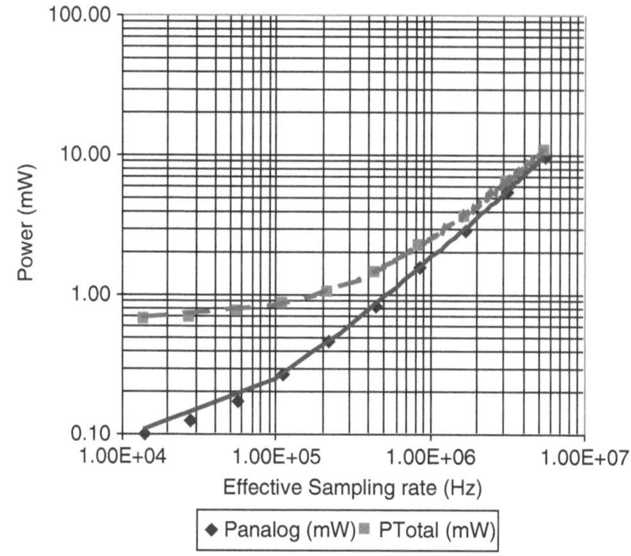

Fig. 7.78 Analog and total ADC power variation with effective sampling rate for $f_{sm} = 30$ MHz

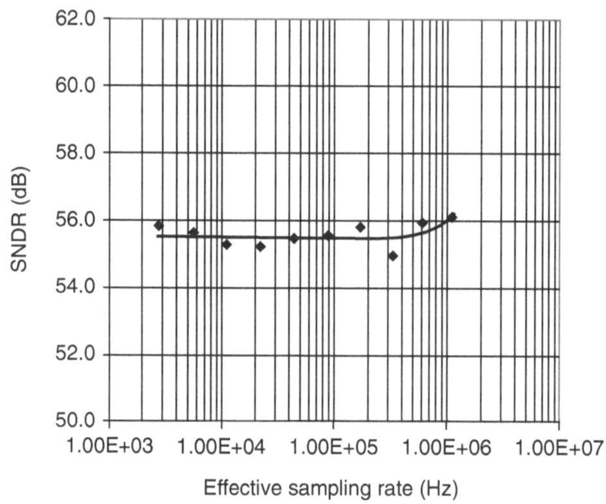

Fig. 7.79 SNDR variation with effective sampling rate for $f_{sm} = 10$ MHz

desired yield and bias point sensitivity required the bias current to be scaled by no more than 1:5 (i.e. 50–10 MHz), the lowest power scalable sampling rate is 2.8 kHz (~1:20,000 × power scalable range). Thus the limiting factor of the CMPS approach is ultimately the limits of transistor performance in weak inversion – however as CMPS multiplies the power scalable range of current scaling by a large

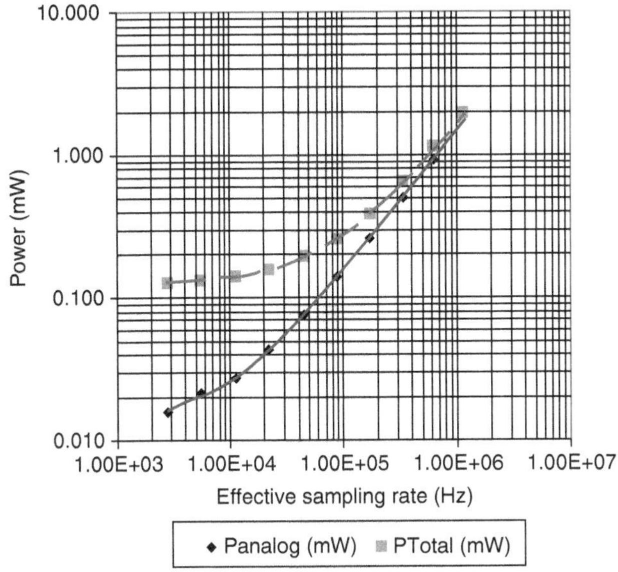

Fig. 7.80 Analog and total ADC power variation with effective sampling rate for $f_{sm} = 10$ MHz

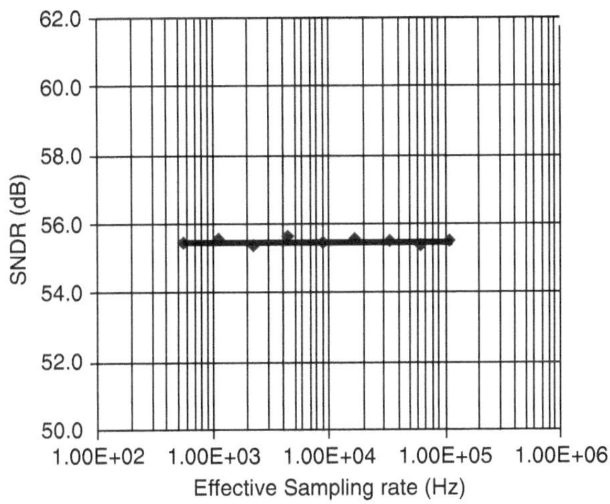

Fig. 7.81 SNDR variation with effective sampling rate for $f_{sm} = 1$ MHz

factor (>1,000 × in the fabricated chip), much lower effective sampling rates with scaled power can be achieved than if the current scaling method were alone used to make power a function of sampling rate.

From the measured results, the ADC requires higher mW per MS/s when using CMPS compared to current scaling. The increased power when using CMPS is only

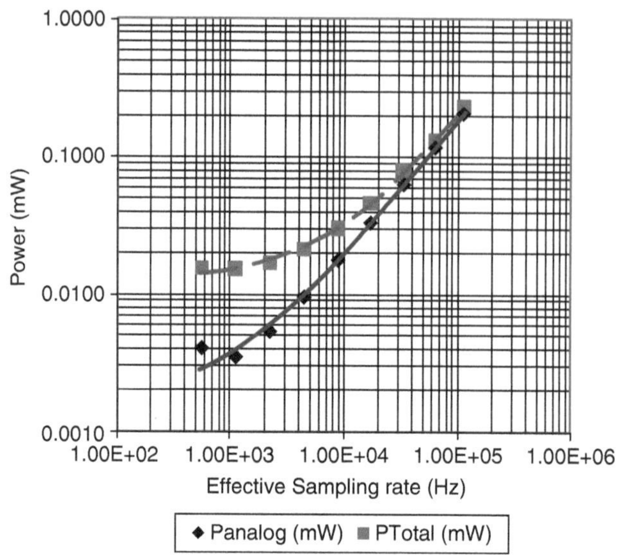

Fig. 7.82 Analog and total ADC power variation with effective sampling rate for $f_{sm} = 1$ MHz

due to a sub-optimal power-triggering scheme. As mentioned in Section 7.6, to simplify the logic of the state machine, all bias circuits were activated at the same time. If the state machine were redesigned to include more programmability, the timing for each bias circuit could be optimized such that the individual bias circuits are powered-on for the minimum time required. As the power of all bias circuits is comparable to single pipeline stage, a significant power reduction can be realized through careful power timing optimization.

The complete power scalable range showing the use of current scaling over a narrow range of sampling rates (1–50 MS/s), and CMPS applied to the narrow range of current scaled sampling rates is shown in Fig. 7.83.

Thus from Fig. 7.83, with CMPS applied to current scaled sampling rates between 1 and 50 MS/s, it is possible to achieve a power scalable power for sampling frequencies as low as 580 Hz (16 μW), and as high as 50 MHz (35 mW).

As the bias currents remain fixed between sampling rates when using CMPS, the advantages of transistors biased in strong inversion can be preserved for low sampling rates, where if current scaling were used, the transistors would be in weak inversion. A specific example displaying strong inversion performance at low sampling rates is the bias sensitivity of the ADC. As shown in Fig. 7.74, for lower sampling rates where current is scaled to achieve lower power for lower sampling rates, the variation of ENOB with bias voltage becomes much larger when the ADC is biased in weak inversion. In Fig. 7.84, the variation of ENOB with bias voltage is shown for $f_s = 1$ MS/s and $f_s = 100$ kS/s when the CMPS is used to power scale (with $f_{sm} = 50$ MHz), and when current scaling is used to power scale.

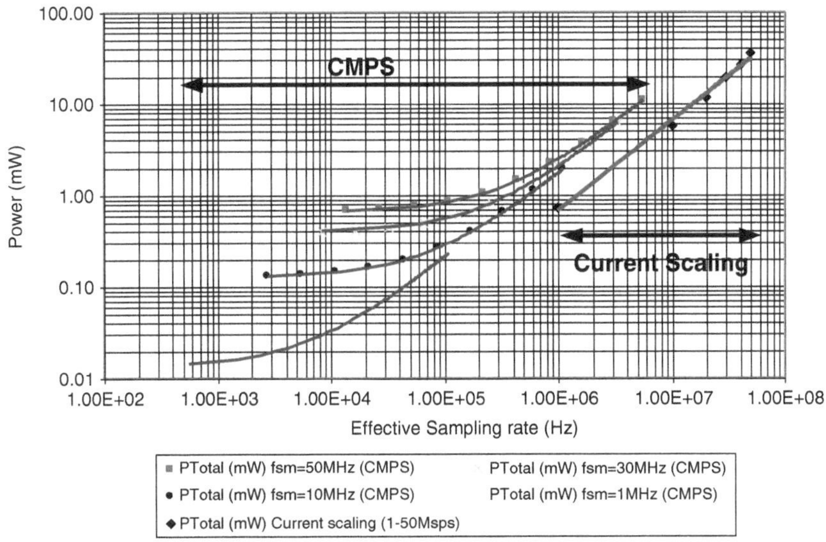

Fig. 7.83 Power scalable range of ADC with CMPS applied to current scaled sampling rates of 1–50 MS/s

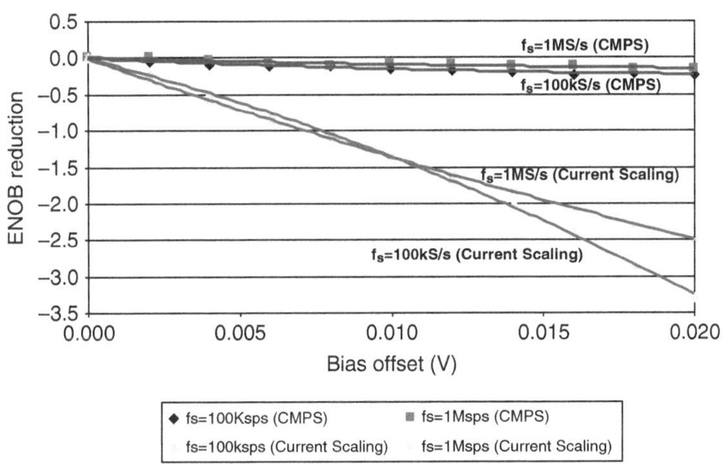

Fig. 7.84 Bias point variation of ADC using CMPS and current scaling for f_s = 1 MS/s, and f_s = 100 kS/s

As expected, the ADC maintains strong inversion performance when CMPS is used, as evident from the small reduction in ENOB with bias voltage variation. Thus for lower sampling rates where CMPS is used, performance and yield degradation associated with weak inversion are minimized.

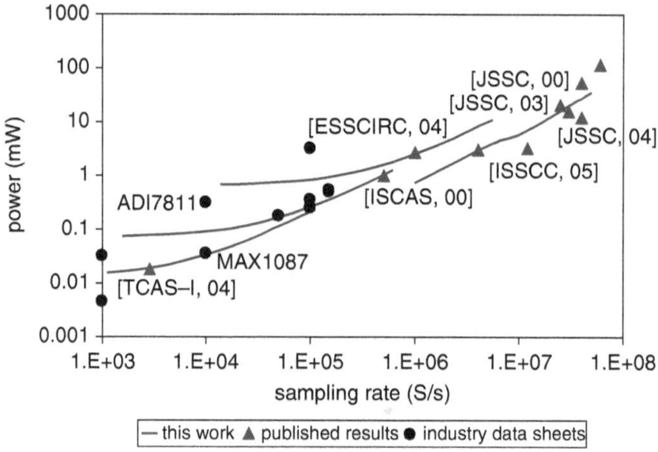

Fig. 7.85 Comparison of power to recently published works and industrial parts

Table 7.11 Summary of key results

Technology	1.8 V, 0.18 μm CMOS
Area	1.21 mm^2
Sampling rate (f_s)	<1 kS/s – 50 MS/s
Input signal swing	1.6 V p-p
Power	15 μW (1 kS/s) – 35 mW (50 MS/s)
SNDR (f_{in} = 20.94 MHz)	55 dB (8.8 bit ENOB)
SFDR (f_{in} = 20.94 MHz)	67 dB
INL/DNL	+0.63/−0.91, + 1.06/−1.2 LSB

The following figure compares the power of the ADC of this work in CMPS and current scaling modes to recently published 10-bit ADCs (Fig. 7.85).

Although the goal of this chapter favored proof-of-concept over absolute figure-of-merit, the ADC in power still exhibits a good figure of merits for a 0.18 μm design – a testament to the benefits gained using the Rapid Power-On Opamp as a means to reduce power in high speed switched capacitor circuits.

7.23 Summary

This chapter presented the design and measured results of a power scalable pipelined ADC fabricated in a 1.8 V, 0.18 μm CMOS process. Measured results showed that at 50 MS/s the ADC consumed 35 mW, and for lower than 1 kS/s only 15 μW was consumed, while maintaining >54 dB of SNDR over the entire power scalable range. The benefits of powering off the MDAC opamps during the sampling phase were also discussed and quantified, where a reduction of power by 20–30% was measured. Increased bias voltage sensitivity as the ADC is driven into weak inversion operation was empirically quantified. The benefit of strong inversion design for low sampling rates using CMPS was shown where bias voltage sensitivity was compared for the ADC using current scaling and CMPS, where in CMPS the ADC shows minimal accuracy degradation with bias voltage fluctuation. Table 7.11 summarizes the results of the ADC presented in this chapter.

Chapter 8
A Sub-sampling ADC with Embedded Sample-and-Hold

8.1 Overview

This chapter builds upon the concepts developed in Chapter 7. In this chapter a 10-bit power scalable ADC with maximum sampling rate of 50 MS/s is also presented, however the ADC of this chapter is targeted for sub-sampled systems where the input frequency can be much higher than the Nyquist sampling rate. To save power, the ADC in this chapter uses a technique to eliminate the front-end sample-and-hold, yet not require carefully matched input paths between the sub-ADC and MDAC of the first stage. The technique allows for a power savings of > 20% compared to the ADC described in Chapter 7. A method to improve the settling behavior of Rapid Power-On Opamps is also presented. Measured results in a 1.8 V, 0.18 μm CMOS process show the embedded sample-and-hold technique to allow for input frequencies higher than 267 MHz. With $f_s = 50$ MS/s, for $f_{in} = 79$ MHz the SNDR was 51.5 dB, and with $f_s = 4.55$ MS/s for $f_{in} = 267$ MHz the SNDR was 52.2 dB.

The organization of the discussion in this chapter is as follows: Section 8.3 describes the technique used to eliminate the front-end sample-and-hold, Section 8.4 describes the circuit implementation of the design, Section 8.7 presents measurement results, and Section 8.8 concludes the chapter.

8.2 Motivation

All mobile communication systems consist of a receive path, in which it is typically required to down-convert a high frequency input down to an IF or baseband frequency. From Section 2.7, it was shown that by using a sub-sampled ADC a front-end mixer could be eliminated from a receive system, and thereby potentially save power. The tradeoff in using a sub-sampled ADC however is that the ADC is required to digitize inputs over a very wide range of frequencies higher than the Nyquist sampling rate. As discussed in Section 3.10 large input frequencies for a

I. Ahmed, *Pipelined ADC Design and Enhancement Techniques*,
Analog Circuits and Signal Processing,
DOI 10.1007/978-90-481-8652-5_8,

pipelined ADC require a power hungry front end S/H to ensure functionality. In the interest of reducing power consumption, techniques to eliminate the front end S/H are developed and presented in this chapter. From Section 5.4.1 it is noted however that the present state of the art techniques to eliminate the front-end S/H rely on carefully matched signal paths and thus a lengthy and costly design procedure. Thus the goal of this work was to develop a sub-sampled ADC topology which enables the elimination of the front-end S/H to save power, yet does not require a carefully matched (thus costly) layout to ensure functionality.

From Chapter 7, power scalability was shown to be greatly useful in allowing a single ADC design to be used for a variety of different standards and operating parameters. As down sampling is used in a variety of different communication systems (especially mobile, low power systems), by expanding the application of power scalability from low pass inputs to band-passed inputs, the ADC of this chapter expands the utility of the power scalable techniques described in Chapter 7.

8.3 Embedded S/H Technique

Figure 8.1 illustrates a conventional 1.5 b stage in a pipelined ADC. During Φ_1 the input is sampled on capacitors C_1 and C_2. During Φ_2 a gain of two is implemented by discharging the charge stored from C_1 to C_2, and DAC operation by connecting V_{DAC} to a voltage set by the sub-ADC.

In this work the first 1.5 b pipeline stage is modified and an additional clock phase introduced as shown in Fig. 8.2. Like the conventional pipeline stage of Fig. 8.1, during Φ_1 the input is sampled on C_1 and C_2. However in this work, when Φ_2 switches from low to high and Φ_{2D} is low, V_{DAC} is set to a high impedance and C_2 is connected between the output and input of the opamp as shown in Fig. 8.3.

Connecting C_2 around the opamp produces a held value of the sampled analog input, and due to charge conservation at the negative input of the opamp the voltage

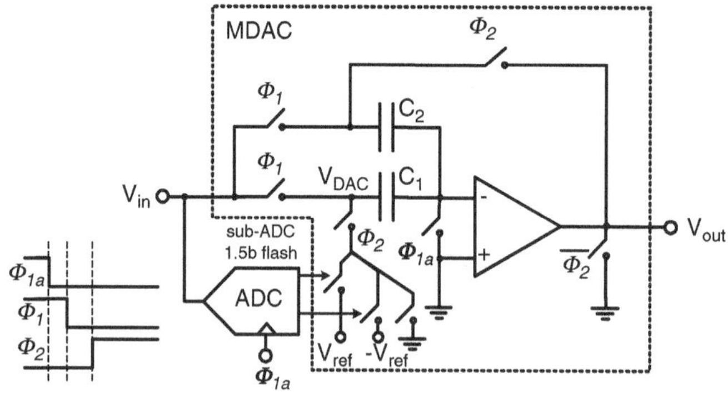

Fig. 8.1 Conventional 1.5 bit MDAC

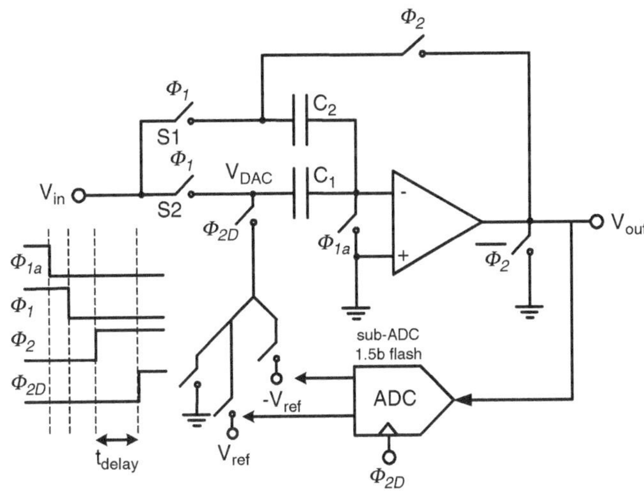

Fig. 8.2 MDAC of this work which enables elimination of front-end S/H (shown single-ended, but implemented fully-differentially)

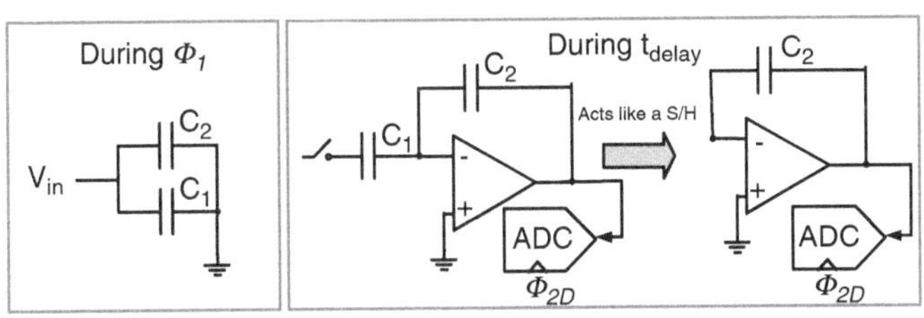

Fig. 8.3 Detailed illustration of MDAC functionality during t_{delay}

across C_1 is preserved as shown in Fig. 8.3. Thus with the approach of this work, during the time labeled t_{delay} in Fig. 8.2, the output of the first stage can be connected to a 1.5 b flash ADC. When Φ_{2D} subsequently goes high the 1.5 b flash ADC resolves its input, sets V_{DAC} to the appropriate DAC voltage, and implements a gain of $\times 2$ by discharging the charge from C_1 into C_2. Therefore the first pipeline stage of this work implements the same functionality as a conventional pipeline stage, however is not sensitive to skew at the input as by using the embedded sample-and-hold of the MDAC, the sub-ADC operates on the same input that is sampled by the MDAC regardless of input frequency. Thus by using the modified first pipeline stage a front-end sample-and-hold is not required to ensure functionality for high input frequencies, hence allowing for substantial power savings.

Since the flash ADC only requires an input that is >1.5-bit accurate, the output of the opamp during t_{delay} is only required to settle to >1.5-bit accuracy to generate the correct outputs from the flash ADC, hence t_{delay} is only a small fraction of the total available settling time as shown in Fig. 8.4.

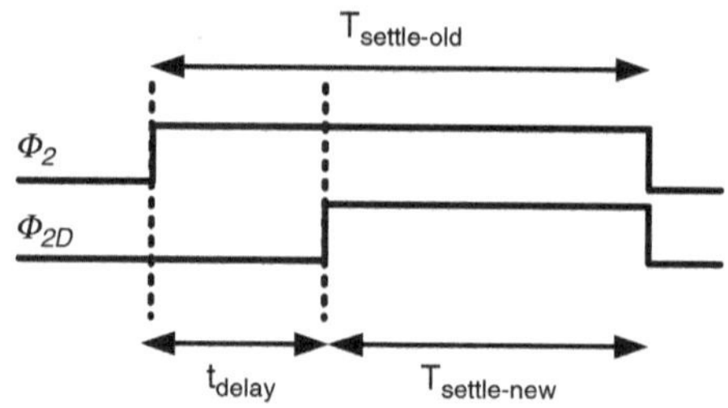

Fig. 8.4 Comparison of MDAC settling time of this work versus conventional MDAC

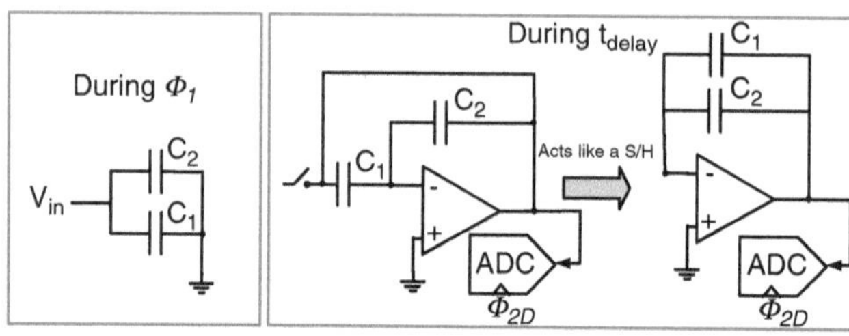

Fig. 8.5 Alternative configuration of MDAC during t_{delay} without floating capacitors

Although the power of the opamp in the first pipeline stage is slightly increased by the fraction of settling time taken by t_{delay}, the overall power of the ADC is significantly reduced as the power hungry front-end S/H is eliminated. The technique to eliminate the front-end S/H could also be applied to multi-bit pipeline stages, noting that prior approaches which relied on redundancy to eliminate the front-end S/H have an even smaller allowable skew thus demand an even more meticulously matched layout than discussed in Section 5.4.1. It should be noted that the technique to use the embedded S/H, although independently derived for this work, is similar to that published in [76]. However the results of this work show: how the embedded S/H technique can be applied to a power scalable architecture, a much higher input bandwidth, and a 66% increase in sampling rate.

From Fig. 8.3 it is noted that during t_{delay} one end of C_1 is floating. Alternatively it is also possible to configure the MDAC such that during t_{delay} node V_{DAC} is also connected to the output of the opamp as shown in Fig. 8.5.

Using the alternative approach the functionality of the MDAC is the same as Fig. 8.2 however a floating capacitor is avoided. By using the approach of Fig. 8.5, the feedback factor of the MDAC during t_{delay} is increased; however the load capacitance seen by the opamp is also increased. Depending on the relative size of C_1 compared to the opamp input capacitance, the approach of Fig. 8.5 could yield a faster or slower solution compared to that in Fig. 8.3. To evaluate both approaches, a programmable switch was included in this work to allow the ADC to operate as shown in Fig. 8.3 or as shown in Fig. 8.5, so that the viability of each approach could be validated.

8.4 Circuit Implementation

The ADC topology of this work is based on the ADC presented in Chapter 7 [4], and thus the ADC of this chapter also has a power scalable architecture. All the circuits from Chapter 7 are reused in this work with the exception that the first stage in this work is implemented using the techniques described in Section 8.3, and no front-end S/H is used in this work. Furthermore in this work a clock delay generator circuit is added, as well as an additional (but not strictly necessary) bias circuit for the first stage to improve noise isolation between the first stage and subsequent pipeline stages. An advantage of basing this work on a previous circuit design is the improvement in power reduction can be made by directly comparing the power consumption of the ADC in this chapter with the power consumption of the ADC in Chapter 7. As the goal of the prototype was proof-of-concept, a specific application was not targeted, however as the ADC is power scalable a variety of applications from low power sensors to wireless LAN could take advantage of the 10-bit architecture.

8.4.1 ADC Architecture

The architecture of the pipelined ADC of this work is shown in Fig. 8.6. The first stage is as shown in Fig. 8.2, and all remaining stages are standard 1.5 b stages. Stages 3–9 are identical to those used in Chapter 7. Stage 2 is a standard 1.5 b stage but has the same sized sampling capacitors and opamp as stage 1.

As there was no explicit front-end S/H block in this work, the total sampling capacitances of the first two pipelined stages in this work were reduced by 40% compared to Chapter 7 (from 940 to 580 fF in stages 1 and 2) while maintaining the same input referred noise as in the ADC of Chapter 7. In this work the opamp of

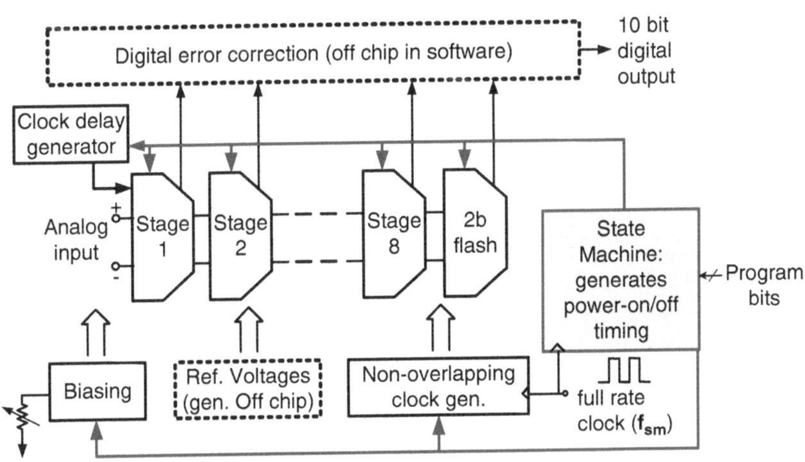

Fig. 8.6 Architecture of pipelined ADC

stage 1 was conservatively designed such that t_{delay} was 20% of the total settling time. However t_{delay} was made tunable so that the lowest t_{delay} possible without gross MSB errors could be measured. The opamps of stage 1 and 2 were made 33% smaller than those used in Chapter 7, yet the opamps had the same settling time due to the smaller capacitive enabled with the techniques presented in this chapter. Thus although less settling time is available in the first pipeline stage, overall power is still reduced without a front-end S/H since the thermal noise floor of the pipeline stages can be made higher (with smaller sampling capacitors) to maintain a fixed input referred noise floor. To enable a large input bandwidth, bootstrapped switches [77] were used as the input switches S1, S2 in Fig. 8.2.

8.4.2 Rapid Power-On Opamp

A Rapid Power-On Opamp based on the design presented in Section 7.8.2 was used in each pipeline stage in this work to realize a power scalable architecture and is shown in Fig. 8.7.

Like conventional gain-boosted opamps, the Rapid Power-On Opamp requires the unity gain frequency of the gain booster opamps (opamps A_p, A_n in Fig. 8.7) to be higher than the 3 dB frequency of the closed-loop but lower than the second pole of the main opamp [21]. Furthermore the loop formed by the gain boosters (labeled L_{Ap} in Fig. 8.7) is also required to be stable. Rapid Power-On Opamps power on and completely off each clock cycle, hence the inputs to the gain booster opamps A_p, A_n effectively see a step function every clock cycle. Thus to ensure good settling, loop L_{Ap} requires a high phase margin. In Chapter 7 standard folded cascode PMOS-input opamps as shown in Fig. 8.8 were used for the gain booster opamps A_p.

The phase margin of the loop L_{Ap} is limited by the second pole of the loop which occurs at the PMOS current mirror node labeled V_{p2} in Fig. 8.8. The same bias voltages used for the main opamp were also used for the gain booster opamps, thus PMOS transistors were sized $\sim\times 5$ larger than the NMOS transistors to maintain a

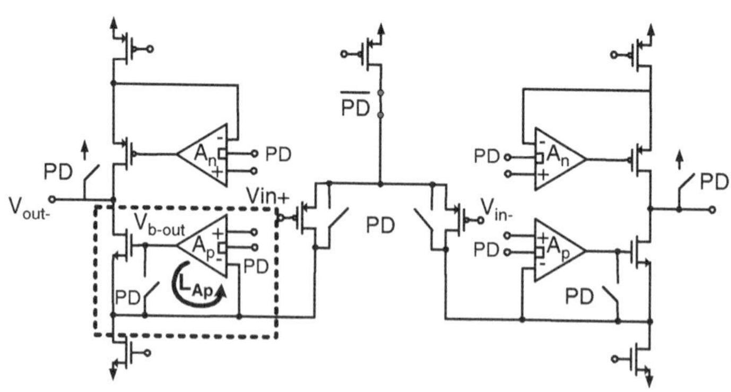

Fig. 8.7 Rapid power-on opamp

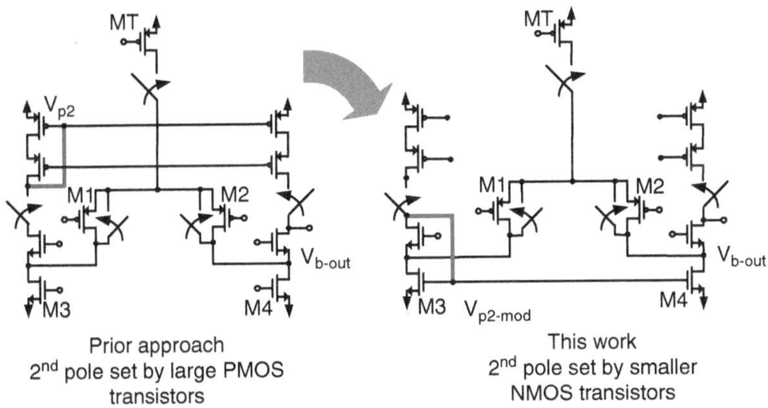

Fig. 8.8 Gain booster opamp used in Chapter 7 (*left*), and in this work (*right*)

similar overdrive voltage as the NMOS transistors for a fixed current density. As such a large parasitic gate-source and gate-drain capacitance exists on node V_{p2} and sets the second pole of the loop to a low frequency which degrades the phase margin of the loop L_{Ap}. To improve the phase margin of loop L_{Ap} the p-input gain booster opamp was modified in this work by implementing the current mirror in the opamp with NMOS transistors instead of PMOS transistors [78] as illustrated in Fig. 8.8. By performing the mirror operation with NMOS transistors, the capacitance on the mirror node is reduced by $\sim\times 2.5$ and the second pole pushed to a much higher frequency – significantly improving the phase margin. Simulation results show that the phase margin of loop L_{Ap} is improved from 56° to 72° by only changing the location of the mirror node. By switching the location of the current mirror node from PMOS to NMOS transistors the slew rate of the gain boosting opamp is reduced by $\times 2$ (i.e. when M2 is cutoff, in the prior approach the current through MT is mirrored by the PMOS current mirror to the output node, however in the approach of this work when M2 is cutoff the output current is set by M4 which is biased with half the current of MT). Although a lower slew rate increases the opamp's power-on time, the significant benefit of a more stable power-on transient makes it a favorable trade-off. Also it is noted that since the power-on time of the opamps of this work only require a small percentage of the settling time to power-on, the total available settling time of the opamp is not significantly reduced with a lower slew rate in the gain booster opamps.

8.4.3 Generation of Delayed Clock Φ_{2D}

The delayed clock edge Φ_{2D} of Fig. 8.4 was generated using a chain of four current starved inverters as shown in Fig. 8.9. Current starved inverters were chosen to allow t_{delay} to be widely tuned for different sampling rates by changing the bias

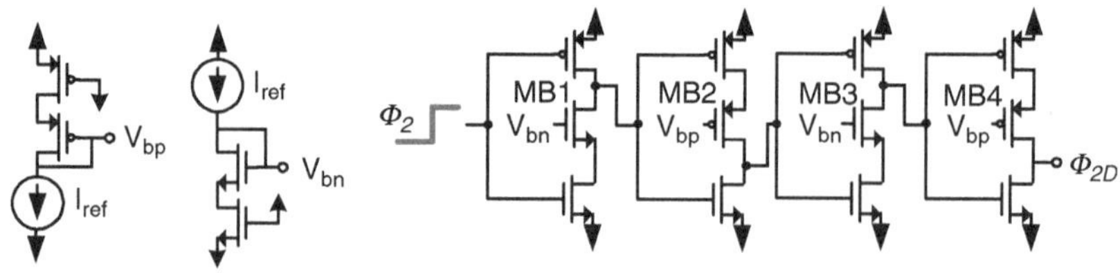

Fig. 8.9 Clock delay block to generate Φ_{2D} from Φ_2

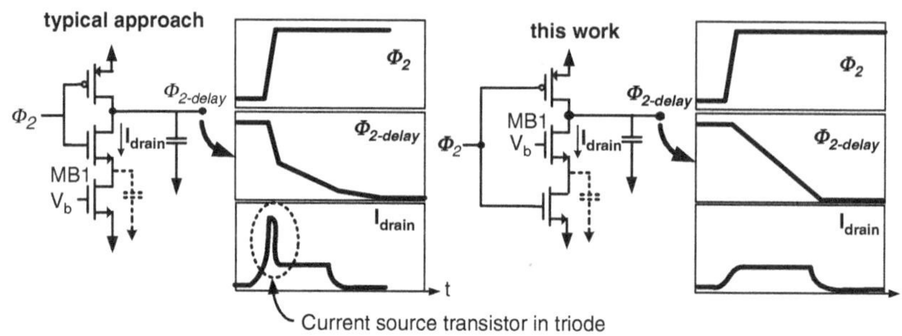

Fig. 8.10 Comparison of different current starved inverter topologies

currents. Also by varying the off-chip reference current I_{ref}, the dependence of ADC SNDR vs. t_{delay}, could be measured.

The current starved inverters were designed such that only one clock edge was delayed. Commonly in a current starved inverter, the current sources are connected to the source nodes of the inverters (e.g. [26, 79, 80]) as shown in Fig. 8.10.

When the input clock switches from low to high in e.g. the first inverter, the current source initially starts off in triode causing a large drain-source current, which is different than the desired bias current, to flow until the pre-charged output discharges to the point where the current source is biased in the active region as shown in Fig. 8.10. As a result the discharge rate of the output also becomes a function of the rise time of the input, reducing the control I_{ref} has on the delay of the current starved inverter, hence reducing range of delay values possible by varying I_{ref}.

In this work, the current source transistors were connected between the drains of the PMOS and NMOS inverter transistors as shown in Fig. 8.10. This was done so that when the input to e.g. the first inverter switches from low to high, the current source transistor MB1 switches from operating in cut-off to the active region, forcing the inverter output to discharge with a rate set by the bias current of MB1. Hence the discharge rate at the output of the inverter is a strong function of the biasing current of MB1 allowing wide variation in the delay of the inverter.

The value of t_{delay} was set by an off-chip reference current I_{ref}. In a practical system for use in industry where a precise definition of the duration of t_{delay} would be required to ensure sufficient settling times for the opamp over process and temperature variations, a DLL can be used to ensure a fixed t_{delay}. Given a sufficient settling time, the clock edge of Φ_{2D} can be allowed to have a large amount of jitter without any significant impact on ADC performance. Hence a potential DLL solution could be implemented with very low power. Furthermore since the most power consuming block of a DLL is typically the delay cell, which is already implemented on-chip in this work, the additional power of a loop filter, charge pump, and phase detector to complete the DLL loop would be relatively small.

8.5 Test Setup: PCB

A four layer FR4 dielectric PCB board with similar layout to that used in Chapter 7, but with some modifications was designed and constructed to test the ADC of this chapter as shown in Fig. 8.11. Separate Power planes were used to isolate the analog, digital, I/O, and board power supplies. A differential input was generated using a 1:1 turns ratio Minicircuits transformer matched to 50 Ω. Reference voltages were generated by passing the output of a resistive voltage divider through an opamp (LM7301) in a unity gain buffer configuration. To maintain constant supply voltages, all voltage supplies for each power plane were generated through regulators (LM337, LM1117), and heavily decoupled with capacitors. As the ADC utilized a constant current biasing scheme, an off-chip adjustable resistor was used as the master current source. The resistance was a series combination of 1, 10 and 200 kΩ, 1 and 3

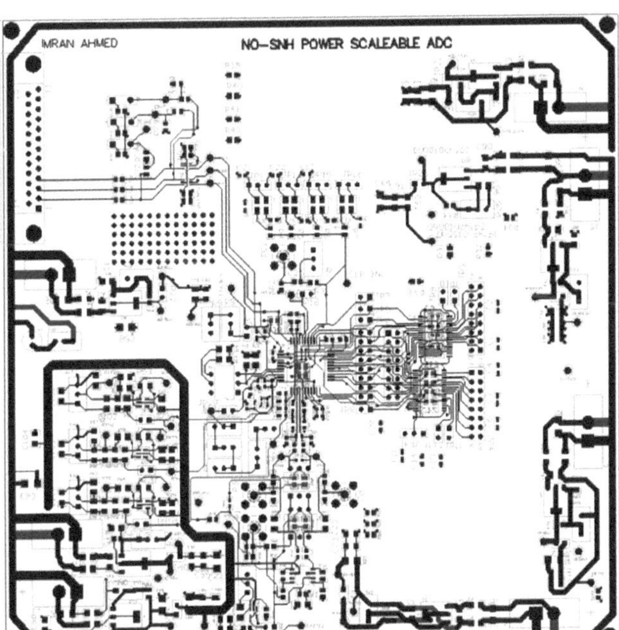

Fig. 8.11 Custom PCB layout

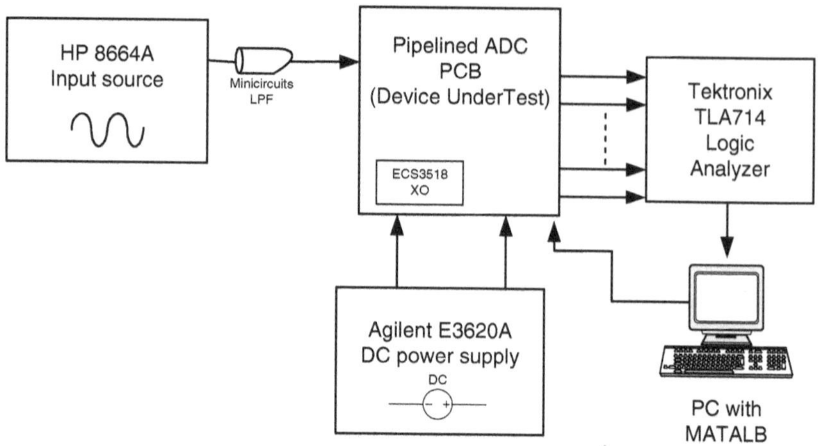

Fig. 8.12 Test setup for pipelined ADC without front-end S/H

MΩ potentiometers such that the biasing current could be accurately controlled over a wide range to facilitate the evaluation of wide range current scaling.

8.6 Test Setup: Equipment

A test setup as shown in Fig. 8.12 was used. Sinusoidal inputs were generated using an HP 8664A function generator. Several Minicircuits filters were used to minimize harmonic distortion from the function generator such that the sinusoidal input to the ADC had an SNDR of well over 62 dB for input frequencies larger than 270 MHz. A 50MHz crystal oscillator (ECS3518-XO) on the PCB was used to generate the low jitter clock for the ADC. The serial shift register was loaded via a parallel port connection to a PC, where a Matlab script was executed to load the appropriate bits. The output bits of each pipeline stage were captured using a Tektronix TLA714 logic analyzer, capable of capturing 65,536 points at a time. An Agilent E3620A Dual output DC power supply was used to provide positive and negative voltages to the voltage regulators on the PCB. The 10-bit output word from the 10-bit ADC was determined via a Matlab script written to emulate the operation of a digital error correction circuit.

8.7 Measured Results

A prototype of the ADC of this work as shown in Fig. 8.13 was fabricated in a 1.8 V, 0.18 μm CMOS process. The core area was 1.1 × 1.1 mm, and the maximum input signal swing was 1.6V p-p differential.

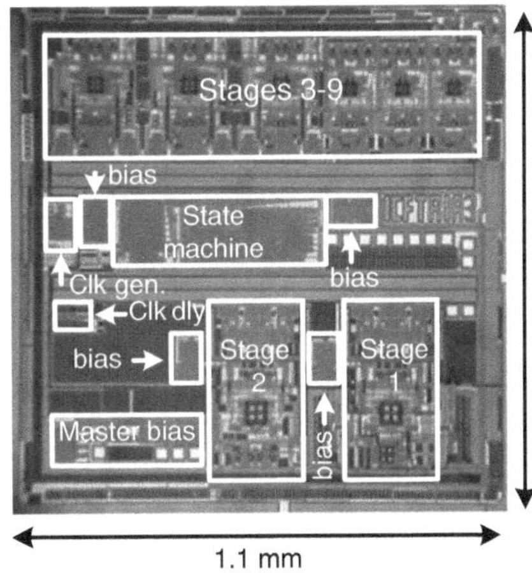

Fig. 8.13 Micrograph of fabricated chip in 1.8 V, 0.18 μm CMOS

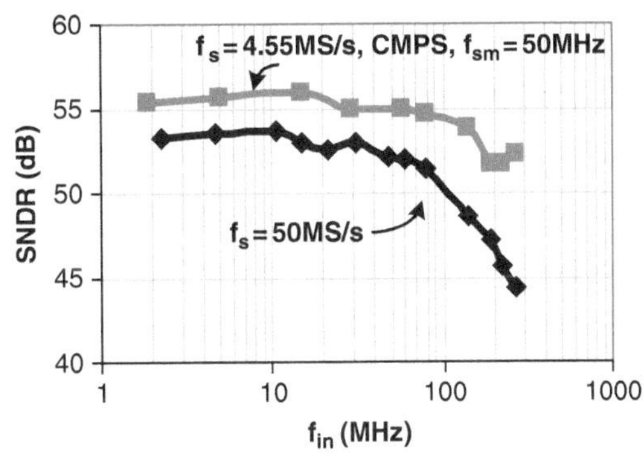

Fig. 8.14 SNDR versus input frequency for $f_s = 50$, 4.55 MS/s

8.7.1 SNDR Versus Input Frequency

Figures 8.14 and 8.15 show the SNDR of the ADC versus input frequency for various power scaled sampling rates. The ADC remains fully functional for input frequencies larger than 267 MHz – frequencies which prior techniques using the redundancy of the first pipeline stage would require a very well matched layout (<<140 ps skew).

For $f_s = 50$ MS/s and $f_{in} = 79$ MHz the SNDR is 51.5 dB. For lower sampling rates the input bandwidth is increased, e.g. for $f_s = 4.55$ MS/s, the SNDR is 52.2 dB for $f_{in} = 267.1$ MHz.

Figure 8.16 shows the FFT of the ADC output for $f_s = 50$ MS/s and 4.55 MS/s. In this work the embedded S/H technique is demonstrated for input frequencies $\sim\times 4$ larger than in [76]. It is noted that the measurements for $f_s = 4.55$ MS/s are for the case when CMPS is enabled such that the settling time in each opamp as well as t_{delay} is the same as the case when $f_s = 50$ MS/s [4]. Also note that although a single sinusoidal input was used, distortion products are captured in the results due to aliasing from sub-sampling.

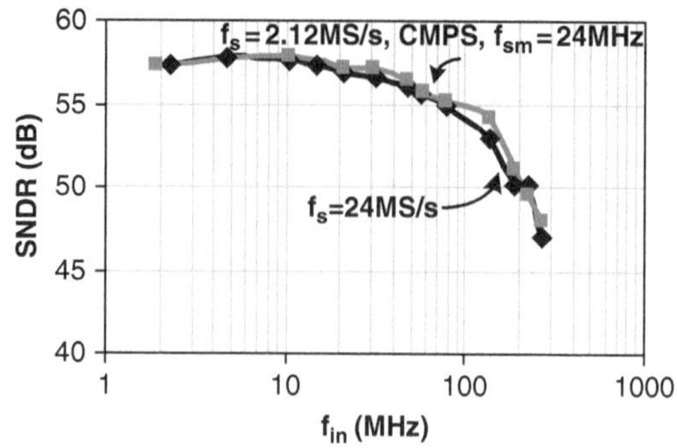

Fig. 8.15 SNDR versus input frequency for $f_s = 24$ MS/s, 2.12 MS/s

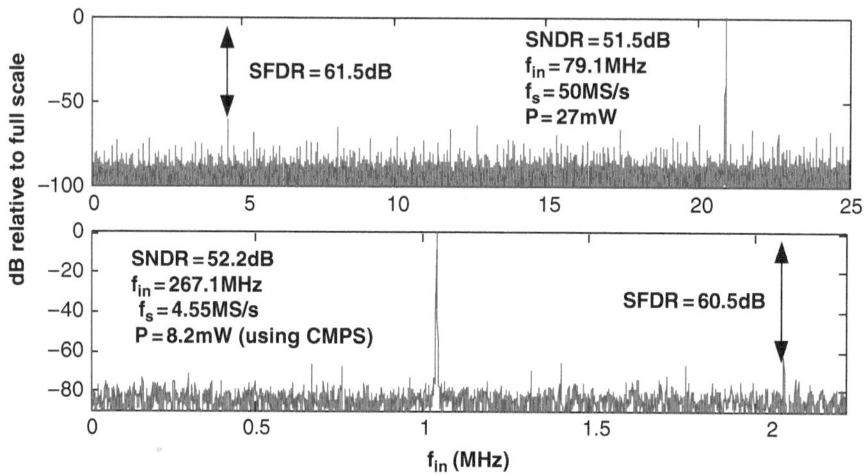

Fig. 8.16 FFT of ADC output at $f_s = 50$, 4.55 MS/s

8.7.2 *Power Versus Sampling Rate*

Figure 8.17 shows the power of the ADC versus sampling rate. The power of the ADC at $f_s = 50$ MS/s was 27 mW, >20% lower than the 35 mW in the ADC of Chapter 7, while maintaining a similar SNDR. It is noted in this work the 27 mW includes an additional bias circuit (1 mW) to improve the robustness of the system, as well as clock delay generator (0.5 mW) to generate Φ_{2D}.

Lower sampling rates with correspondingly lower power can be realized by increasing t_{ON} while current scaling the ADC as was demonstrated in Chapter 7. From Fig. 8.17 the power of the ADC of this work is shown to compare favorably to other recently published 10-bit ADCs [4, 61, 76, 81–92] over a wide range of sampling rates.

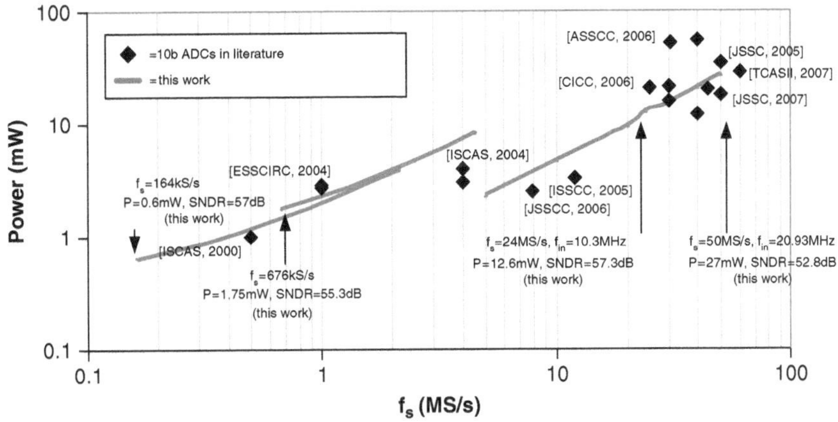

Fig. 8.17 Power versus sampling rate

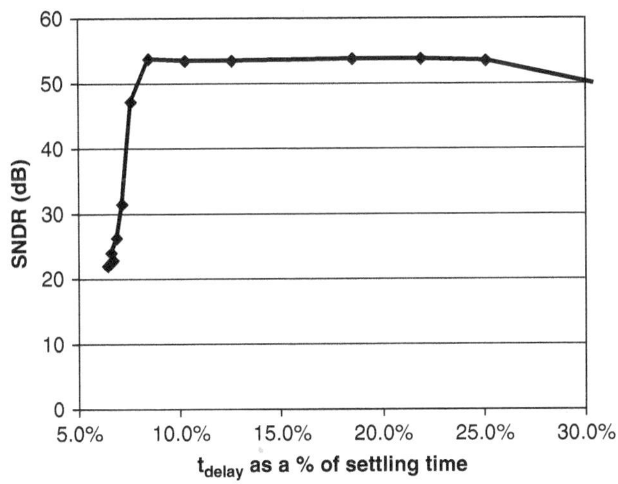

Fig. 8.18 Variation of ENOB with t_{delay} using approach of Fig. 8.3

8.7.3 T_{delay} Versus Settling Time: Robustness of Technique

In Figs. 8.18 and 8.19 the SNDR of the ADC is plotted versus the percentage t_{delay} is of the total available settling time when the first pipeline stage is configured using the approach of Figs. 8.3 and 8.5 respectively when $f_s = 50$ MS/s.

From both plots it can be concluded that MSB errors only occur when t_{delay} is $<10\%$ of the available settling time. Thus the technique to use the embedded sample-and-hold described in this work does not require the first stage opamp to be significantly increased in power to maintain settling accuracy. Furthermore the

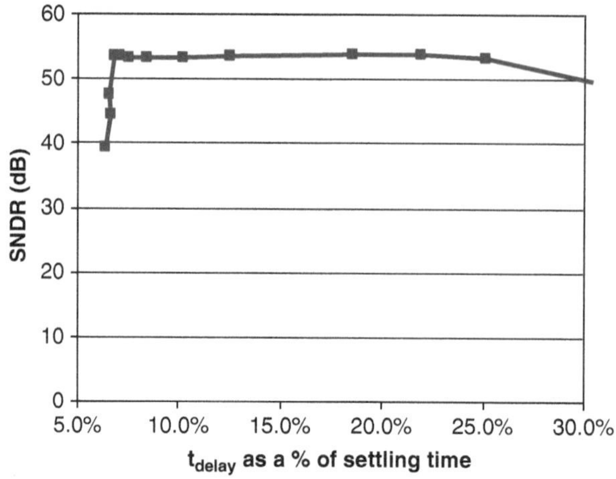

Fig. 8.19 Variation of ENOB with t_{delay} using approach of Fig. 8.5

Table 8.1 Summary of key results

Technology	1.8 V, 0.18 μm CMOS
Area	1.21 mm^2
Sampling rate (f_s)	<164 kS/s–50 MS/s
Input signal swing	1.6V p-p differential
Power	<0.6–27 mW
SNDR	>51.5 dB for all f_s
SFDR	>60.5 dB for all f_s
Input frequency range	0–267 MHz
Chapter 7 ADC power @ 50 MS/s	35 mW
Power of this work at 50 MS/s	27 mW

fact that the SNDR only degrades for t_{delay} larger than 25% of the settling time indicates the power of the first stage opamp could easily be further reduced. Both approaches exhibit similar performance for different t_{delay}, where the approach of Fig. 8.5 shows better performance for slightly smaller values of t_{delay}. This can be attributed to the fact that although the load capacitance of the opamp in the first stage is increased using the approach of Fig. 8.5, for the opamp used in the first stage of this work, the larger feedback factor ultimately allows for faster settling.

8.8 Summary

A power scalable pipelined ADC for use in sub-sampled systems with a large input bandwidth was described. Using a technique to remove an explicit front-end sample-and-hold block, a power savings of >20% compared to the ADC of Chapter

7 was realized. A method to improve the settling behavior of Rapid Power-On Opamps was also presented. Measured results from a 1.8 V, 0.18 μm CMOS prototype show the ADC to achieve more than 51 dB SNDR for input frequencies larger than 79 MHz for $f_s = 50$ MS/s and 267 MHz for $f_s = 4.55$ MS/s. Table 8.1 summarizes key performance metrics of the results presented in this chapter.

Chapter 9
A Capacitive Charge Pump Based Low Power Pipelined ADC

9.1 Overview

In this chapter a pipelined topology which significantly reduces ADC power consumption and does not require opamps is presented [8]. The gain of each pipeline stage in this work is set by a technique which allows for a stage-gain close to $\times 2$ to be achieved passively and differentially, where digital foreground calibration is used to compensate for linear gain errors in each pipeline stage. The only active circuit in each pipeline stage is a unity gain buffer which drives the subsequent pipeline stage. In this work, the unity gain buffer was implemented with an NMOS source follower, which was designed to have sufficient linearity to meet the desired 10-bit requirements of this work for a signal swing of 1 V p-p. The open-loop technique of this work allows for power savings on the order of those presented in [74], however has a much simpler calibration architecture as non-linearities from the gain stage are not required to be calibrated. Furthermore unlike [75], the approach of this work allows for great power savings but with a topology which is differential. Measured results from a prototype fabricated in a 1.8 V, 0.18 um CMOS process show the 50 MS/s ADC to achieve a peak SNDR/SFDR of 58.2/66 dB after calibration, where the power consumption of the ADC was only 9.9 mW for a figure of merit of 0.3 pJ/step.

Section 9.2 discusses the motivation of the low power ADC. Sections 9.3–9.5 discuss the implementation of the novel low power ADC. Section 9.6 analyzes the thermal noise of the approach discussed in this chapter. Section 9.7 reviews the foreground calibration technique used in this chapter. Section 9.8 presents a theoretical analysis of the improvement afforded with the approach of this chapter. Section 9.9 outlines the design specifications of this chapter, and Section 9.10 discusses the circuit implementation of the ADC. Sections 9.11–9.12 detail the test setup and measurement results of a prototype fabricated in a 1.8 V, 0.18 μm CMOS process. Section 9.13 concludes the chapter.

I. Ahmed, *Pipelined ADC Design and Enhancement Techniques*,
Analog Circuits and Signal Processing,
DOI 10.1007/978-90-481-8652-5_9, © Springer Science+Business Media B.V. 2010

9.2 Motivation

The primary motivation of the work presented in this chapter is to significantly reduce power consumption in pipelined ADCs. ADC power reduction enables longer battery life in mobile applications, and lower cost packaging in wired applications. In Section 3.3, it was shown that the opamp (which is used to achieve a linear stage-gain) was the dominant power consumer in pipelined ADCs. As a result, the majority of prior published reports on power optimization techniques in pipelined ADCs ultimately involve power optimizations in the opamp. Since the opamp is the limiting factor in MDAC power consumption, the goal of this work [8] is to fundamentally rethink the design of the MDAC and investigate alternative topologies which do not require power hungry opamps to achieve the desired stage-gain.

9.3 Architecture: Capacitive Charge Pump Based Gain

In this work [8], rather than obtaining stage-gain from capacitors in a negative feedback network around an opamp, a capacitive charge pump inspired technique is used to obtain a $\times 2$ gain for a 1.5-bit pipeline stage. Capacitive charge pumps attain voltage gain by sampling an input voltage on multiple capacitors, and subsequently connecting each capacitor in series [93] to yield a total voltage which is the sum of the voltages sampled on each capacitor (e.g. Dickson Charge pump [94]). Figure 9.1 illustrates an example of a voltage doubler using a capacitive charge pump.

From Section 3.2 it is recalled that the gain of a 1.5-bit pipeline stage is $\times 2$, thus it is conceivable that a gain technique similar to that shown in Fig. 9.1 can be employed to yield the stage-gain in a pipelined ADC stage. Figure 9.2 illustrates a potential pipeline stage topology which uses a capacitive charge pump to achieve a stage gain of $\times 2$.

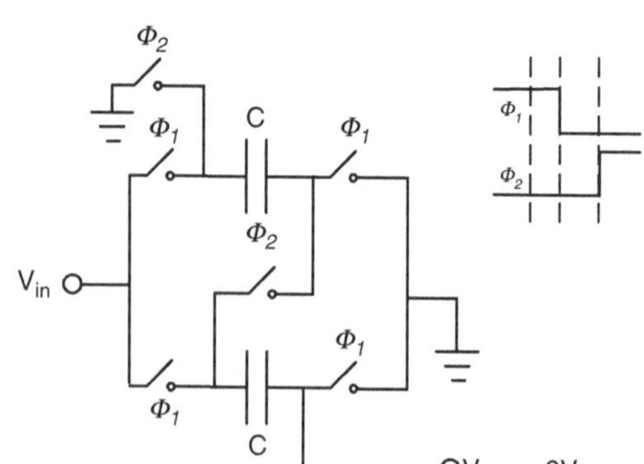

Fig. 9.1 Example of a voltage doubler

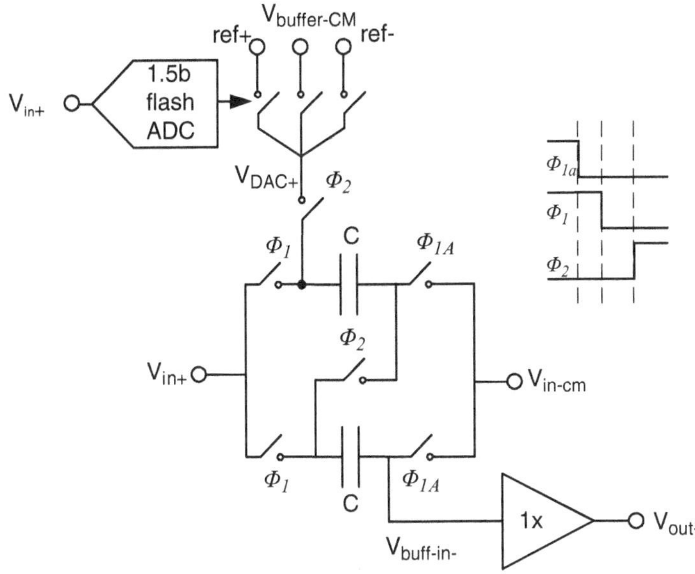

Fig. 9.2 Pipeline stage using a capacitive charge pump

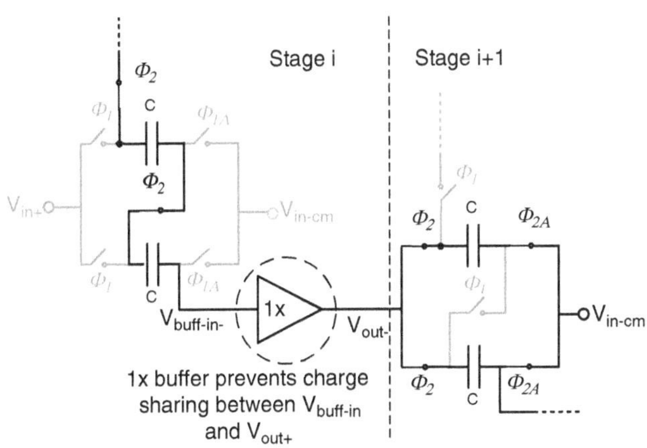

Fig. 9.3 Illustration of how 1× buffer prevents charge sharing

Assuming the subsequent pipeline stage is constructed similarly to Fig. 9.2, the unity gain buffer is added so that charge sharing does not take place between capacitors from one pipeline stage to another as shown in Fig. 9.3.

One of the drawbacks of the approach of the pipeline stage shown in Fig. 9.2 is that there is no common-mode rejection for a differential input signal. For example consider the case where each input V_{in+}, V_{in-} is offset from the desired common

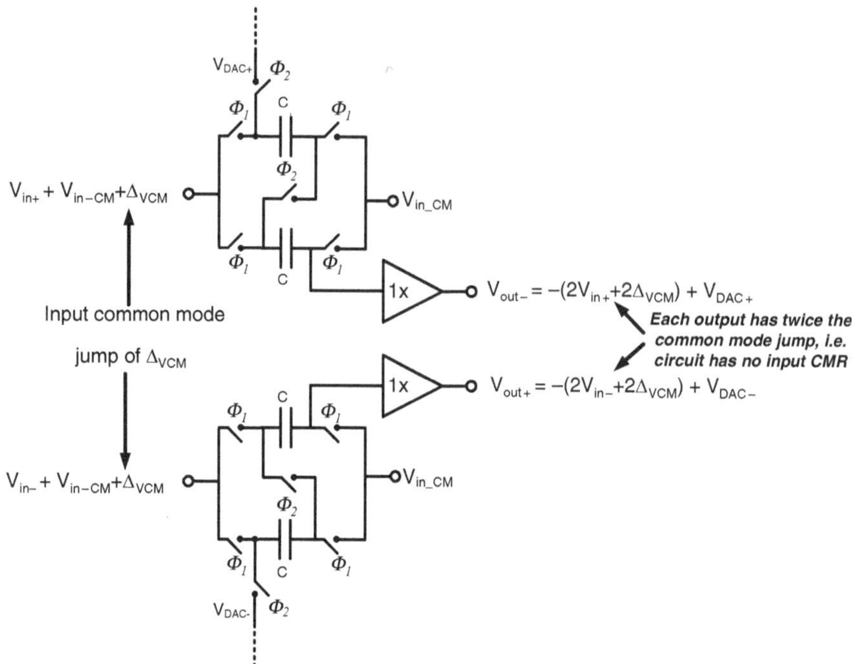

Fig. 9.4 Illustration of poor input CMR for pipeline stage shown in Fig. 9.2

mode voltage by Δ. As shown in Fig. 9.4 this results in the common mode of the output also being doubled in addition to the desired analog input, i.e. the topology of Fig. 9.4 is pseudo-differential. Thus the pipeline stage shown in Fig. 9.2 has very poor input Common Mode Rejection (CMR)

In a pipelined ADC consisting of many stages it is quite likely that with such a structure a small common mode offset at the input of one of the pipeline stages could quickly multiply along the pipeline, making the absolute voltage of the input to a latter pipeline stage approach the supply voltage, thus eliminating the signal swing and hence rendering subsequent pipeline stages unusable. To avoid the poor common mode rejection problem in the architecture of Fig. 9.2, an alternative sampling topology was developed for this work as shown in Fig. 9.5 (and is the topology used to implement each pipelined stage in the ADC of this chapter).

In the topology of Fig. 9.5, since the ***differential*** input is effectively sampled across the series combination of the two sampling capacitors, variations in the common mode voltage are not captured by the sampling capacitors. Switch S0 is included to isolate nodes V_{X1} and V_{X2} during Φ_1, and thus ensure that switches S1 and S2 act as bottom plate sampling switches so as achieve a high linearity by minimizing charge injection effects. Simulation results verified that the sampling

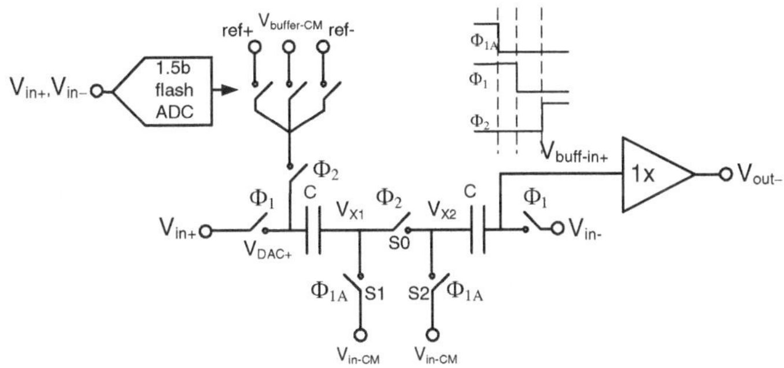

Fig. 9.5 Pipeline stage used in the work of this chapter

topology could easily meet the >10-bit linearity target of this work when the switches were sized large enough to ensure the ADC operated at 50 MS/s, and the sampling capacitors sized large enough to place the thermal noise floor beyond the 10-bit level.

In the MDAC, since the gain of the pipeline stage is derived using a capacitive charge pump technique, the gain of the pipeline stage will not be precisely set by only the ratio of the sampling capacitors. As will be seen in the following Section, the gain is a function of parasitic capacitors in addition to the sampling capacitors. As noted in Section 5.2 however, calibration can be used to measure and compensate for the effect of a non-ideal stage gain, and is thus used in this work. Furthermore, as will be seen in the subsequent sections, using the topology of Fig. 9.5, the stage-gain can be designed to be linear enough that only a simple linear calibration technique (i.e. stage-gain error correction) is required to improve the accuracy of the ADC to the target resolution of this work (10 bits). This is contrasted to the open-loop approach of [74] where by virtue of the gain being derived from a differential pair (which has a non-linear input/output characteristic), a complicated non-linear calibration scheme was required.

The significant advantage of the capacitive charge pump technique is the only active circuitry required is a unity gain block. In general all *linear* unity gain buffers use some form of internal feedback, where since the closed-loop gain of the buffer is $\times 1$ the value of $1/\beta = 1$. Recall from Section 3.5 that the unity gain frequency of a closed-loop system is given by: $\omega_t = \beta\omega_{ta}$, and that the value of β is at best 0.5 for the traditional MDAC shown in Fig. 8.1. Since ω_{ta} is linearly related to the power of the amplifier [10], it can be deduced that the MDAC topology of this chapter can achieve the same speed as the traditional MDAC with half the power consumption, due to the larger value of β in the MDAC of this chapter. In practice it is noted that in high speed ADCs the traditional MDAC suffers from a β smaller than 0.5 due to large parasitic capacitors. Thus the MDAC of this chapter in practice can achieve the same speed as the traditional MDAC, but with *less than half* the power consumption.

An additional advantage of the technique discussed this chapter is: since the buffer comes after the passive gain block, the power of the buffer's noise when referred to the input of the pipeline stage is divided by the amount of passive gain squared – unlike the traditional MDAC where the noise of the opamp is not divided by the stage gain when referred to the input. Thus noise from active circuitry in the topology contributes less to the overall noise floor than in the traditional MDAC, hence enabling further reduction in ADC power.

9.4 Effect of Parasitic Capacitors

In this Section the effect of all parasitic capacitors on the gain of the pipeline stage of Fig. 9.5 is examined, where Fig. 9.6 illustrates the key parasitic capacitors.

From Fig. 9.6, during Φ_1 when a differential input of V_i with a common-mode offset of Δ is sampled, the charge sampled on node V_x during Φ_1 is given approximately by:

$$Q_{\Phi 1\_V_x} = 2\Delta C \tag{9.1}$$

During Φ_2 the total charge on node V_x is given approximately by:

$$Q_{\Phi 2\_V_x} = C_{p-SW} V_X + C(V_x - V_{DAC+}) + C(V_X - V_{buff-}) \tag{9.2}$$

As charge is conserved at node V_x, equations (9.1 and 9.2) can be equated, yielding:

$$V_x = V_{buff-}\left(\frac{C}{2C + C_{p-SW}}\right) + V_{DAC+}\left(\frac{C}{2C + C_{p-SW}}\right) - 2\Delta\left(\frac{C}{2C + C_{p-SW}}\right) \tag{9.3}$$

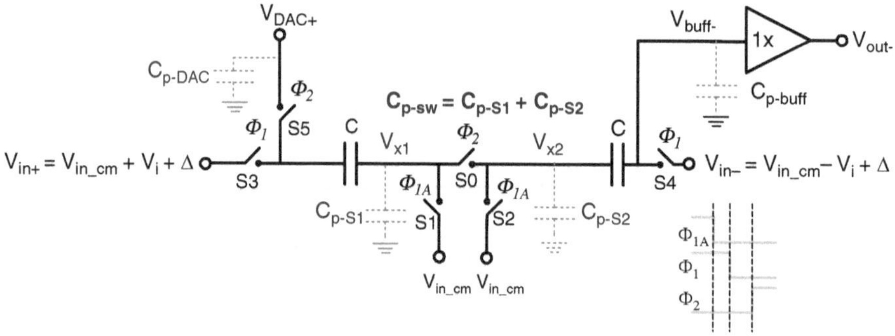

Fig. 9.6 Illustration of parasitic capacitors in MDAC of this chapter

During Φ_2, the charge sampled on V_{buff-} is given by:

$$Q_{\Phi 2\_V_{buff-}} = C_{p-buff} V_{buff-} + C(V_{buff-} - V_x) \tag{9.4}$$

Using the fact that charge is conserved at node V_{buff-} between Φ_1 and Φ_2 gives the following:

$$C(-V_i + \Delta) = C_{p-buff} V_{buff-} + C(V_{buff-} - V_x) \tag{9.5}$$

Substituting (9.3) into (9.5), the expression for the output voltage, V_{out-}, during Φ_2, is given by:

$$\begin{aligned} V_{out-} = -A_{buff} \Bigg[& V_i \frac{2 + \frac{C_{p-SW}}{C}}{1 + \frac{C_{p-SW}}{C} + 2\frac{C_{p-buff}}{C} \frac{C_{p-sw}C_{p-buff}}{C^2}} \\ & - V_{DAC+} \frac{1}{1 + \frac{C_{p-SW}}{C} + 2\frac{C_{p-buff}}{C} \frac{C_{p-sw}C_{p-buff}}{C^2}} \\ & - \Delta \frac{\frac{C_{p-SW}}{C}}{1 + \frac{C_{p-SW}}{C} + 2\frac{C_{p-buff}}{C} \frac{C_{p-sw}C_{p-buff}}{C^2}} \Bigg] \end{aligned} \tag{9.6}$$

where A_{buff} is the gain of the unity gain buffer (which is $\sim\times 1$). When the parasitic capacitors are zero, and $A_{buff} = 1$:

$$V_{out-} = -[2V_i - V_{DAC+}] \tag{9.7}$$

which is precisely the residue transfer characteristic of a 1.5-bit pipeline stage. Assuming $C_{p\,=\,sw} << C$, equation (9.6) shows that a common-mode error of Δ in the input is significantly attenuated in the MDAC's output. It is noted that since node V_{DAC+} is driven by the DAC voltage during Φ_2, parasitic capacitors at node V_{DAC+} have no impact, and thus the size of switch SDAC can be made large without impacting the gain or DAC voltage used in the pipeline stage. However as is clear from equation (9.6) parasitic capacitors at nodes V_{X1}, V_{X2} and $V_{buff\text{-}in\text{-}}$ do affect the gain of the MDAC as well as the DAC voltage subtracted in the output. As noted earlier, calibration is used to compensate for the effect of an MDAC gain which is not exactly equal to 2 as well as the fact that the DAC voltage subtracted is not exactly V_{DAC+}. Furthermore it is noted that that switches S1 and S2 can be sized relatively small (hence have small parasitic capacitance) as they connect to constant DC voltage sources. What is of concern however is the effect of non-linearity in the parasitic capacitors $C_{p\text{-}S1}$ and $C_{p\text{-}buff}$.

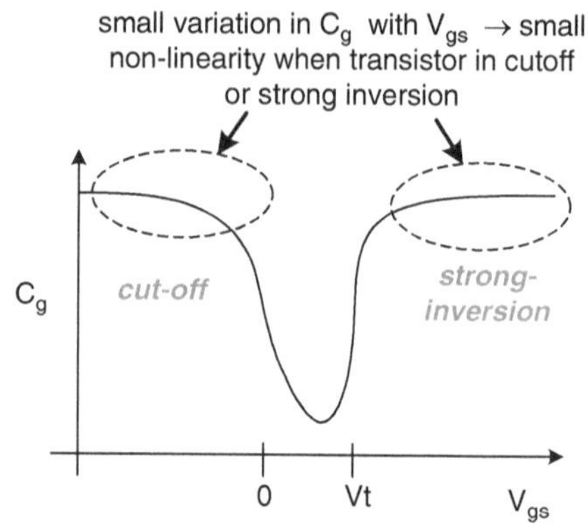

Fig. 9.7 Variation of gate capacitance with gate-source voltage

It is well known that the total gate capacitance of a MOS transistor varies with gate-source voltage with a shape as shown in Fig. 9.7.

From Fig. 9.7, if it can be guaranteed that the gate-source voltage of the switches and input transistor in the buffer are in strong inversion or cut-off during Φ_2, the amount of variation in gate capacitance (which is the dominant parasitic capacitance) with signal swing (i.e. gate and/or source voltages on transistors) can be significantly reduced. Since all switches in the switched capacitor circuit are designed to be strongly inverted to minimize 'on' resistance, or in cut-off when the switch is off, and the input transistor of the buffer also designed to be strong inversion, the impact of non-linearities from parasitic capacitor variation with signal swing is relatively small in this work. Simulation results showed that more than 10-bit linearity could be achieved at 50 MS/s using the topology of this chapter. It is noted however that the effect of capacitor non-linearity could be a limiting factor in preventing the topology from being useful in highly linear ADCs (i.e. >> 10-bit linear).

9.5 Unity Gain Buffer Topology

One of the key advantages of the pipelined topology of this chapter is that the only active circuitry required is a circuit which has a gain of only $\times 1$. From Section 3.5 it was shown that in feedback systems the unity gain frequency of the closed-loop was maximized for a large feedback factor β. Thus the MDAC topology allows one to achieve a gain larger than one, yet has the benefits of active circuitry which only requires a gain of $\times 1$. Furthermore it is noted that since in many linear unity gain buffers the feedback network can be implemented using a simple wire, β truly equal to one can be achieved – unlike capacitive feedback based systems (e.g. traditional MDAC topology using opamps), where parasitic capacitors reduce the value of β.

To determine the best buffer topology for the MDAC of this chapter, several commonly used unity gain buffer stages were surveyed. The following figures and tables summarize the benefits and disadvantages of each buffer surveyed. Of note – each buffer topology is shown such that the total current is sufficient for each topology to have the same unity gain frequency of ω_t, where it is assumed each buffer has the same load capacitance C_L.(Fig. 9.8–9.14)

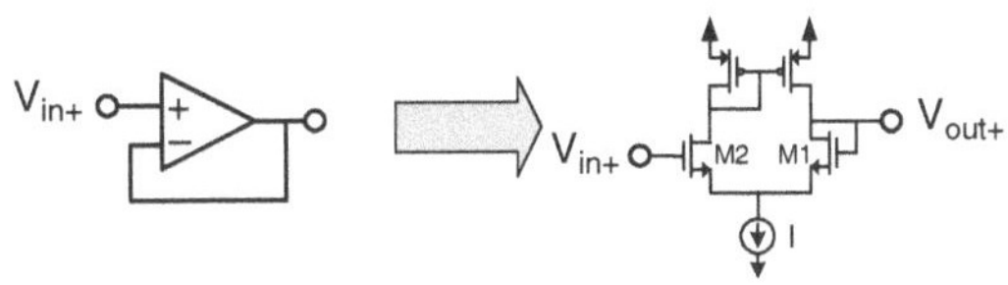

Fig. 9.8 Opamp in unity gain configuration

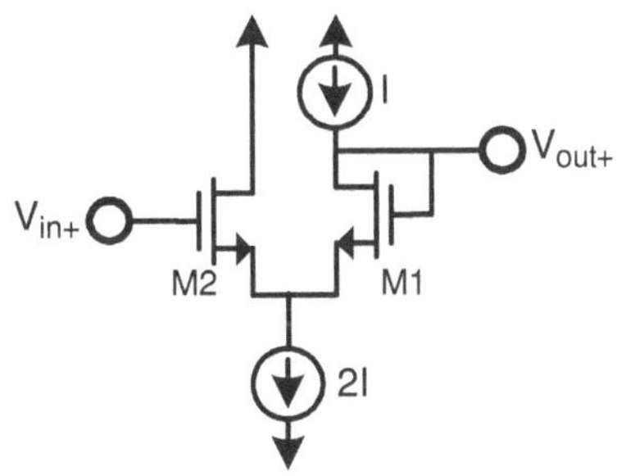

Fig. 9.9 Compound source follower

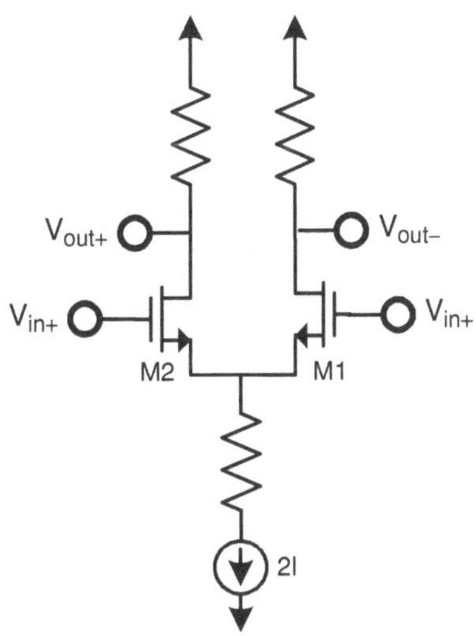

Fig. 9.10 Resistively degenerated differential pair

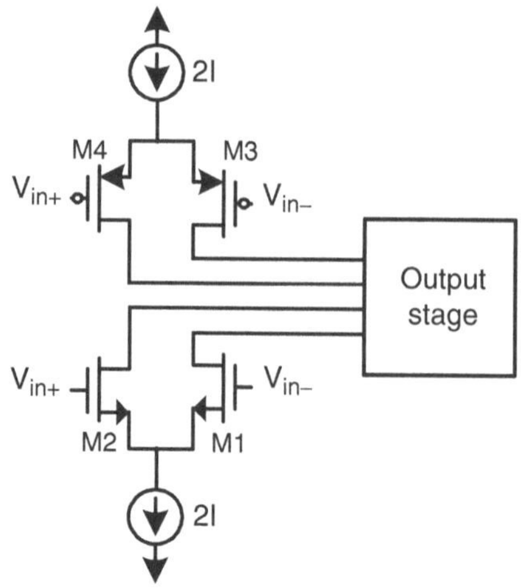

Fig. 9.11 Unity gain buffer which has an N-P complimentary input stage

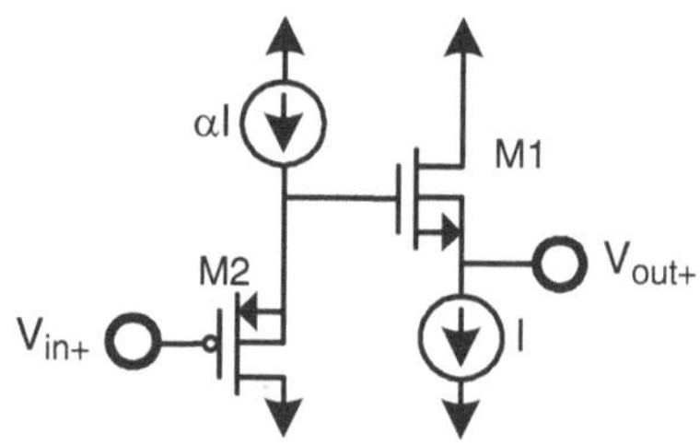

Fig. 9.12 Cascade source follower

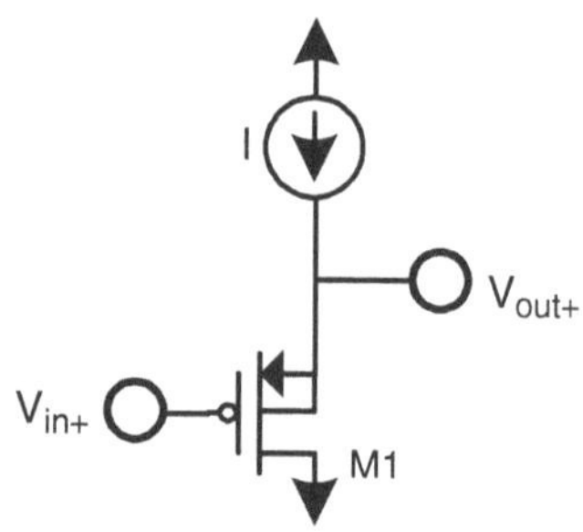

Fig. 9.13 PMOS source follower

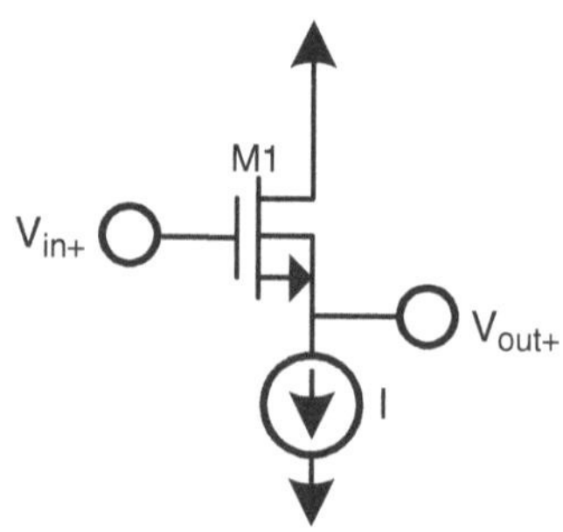

Fig. 9.14 NMOS source follower

Opamp in unity gain configuration	
Advantages	Disadvantages
√ $1/\beta = 1$	x Noise of M1 and M2 contribute to single-ended output noise
√ NMOS diff pair → large g_m	
√ Body effect does not have a huge impact	
√ Output CM tracks input CM	
√ Simulations show > 10-bit linearity for 1V-pp input	
√ Power = IV_{DD}	

Compound source follower	
Advantages	Disadvantages
√ NMOS diff pair → large g_m	x Noise of M1 and M2 contribute to single-ended output noise
√ Body effect does not have a huge	x Power = $2IV_{DD}$
√ Impact	
√ Output CM approx tracks input CM	
Simulations show > 10-bit linearity for 1V-pp input	

Resistively degenerated differential pair	
Advantages	Disadvantages
√ NMOS diff pair → large g_m	x Degeneration reduces bandwidth
√ Can use CMFB to guarantee output CM	x Sensitive to process variation
√ Fully-differential – Input CM rejection is high	Simulations show <10-bit linearity for 1V-pp input
√ Only noise of single transistor M1 or M2 contributes to single-ended output noise	
√ Power = IV_{DD} (single-ended)	

Unity gain buffer which has an N-P complimentary input stage	
Advantages	Disadvantages
√ Can have large input/output CM variation	x g_m Varies with signal swing
√ Larger swing, smaller sampling caps → lower power	x Input offset varies as a function of input swing, leading to poor linearity
	x Simulations including offsets on reasonably sized input differential pairs show < 10-bit linearity
	x Power > IV_{DD}
	x Large noise

Cascade source follower	
Advantages	Disadvantages
√ Output common mode close to input common mode	x Output CM varies over process corners
√ Simple, compact design	x Less headroom
	x Gain < 1 (cascade of two stages with gain < 1)
	x Need DNW
	x Power = $(1 + \alpha)IV_{DD}$
	x Large noise

PMOS source follower	
Advantages	Disadvantages
√ Simple, compact design	x Small g_m with PMOS input
√ Body effect can be easily eliminated	x Output CM not equal to input CM
√ Can get 10 b linearity with reasonable signal swing	x Smaller output signal swing
√ Power = IV_{DD}	
√ Single transistor – low noise	

NMOS source follower	
Advantages	Disadvantages
√ Simple, compact design	x Output CM not equal to input CM
√ NMOS → high gm	x Need DNW to eliminate body effect
√ Can get 10 b linearity with reasonable signal swing	x Smaller output signal swing
√ Single transistor – low noise	
√ Power = IV_{DD}	

From the survey of unity gain buffers the most power efficient and simple buffer structure which meets the minimum 10-bit linearity requirements of this work is the NMOS source follower. The NMOS source follower had the lowest input referred noise power spectral density amongst all surveyed amplifiers, and furthermore by using an NMOS device the g_m of the amplifier is maximized (noting that PMOS transistors usually have one fourth the g_m of an NMOS).

One of the disadvantages of source followers is the different common-mode voltage between input and output nodes of the amplifier. However as noted in Section 9.3, using the topology of this chapter, the common mode at the input of the unity gain buffer is decoupled from the common mode at the output of the unity gain buffer. That is, since the input is differentially sampled on the capacitors, the common mode of node voltage $V_{buff\text{-}in\text{-}}$ in Fig. 9.5 is set approximately by the common mode voltage of the DAC voltage V_{DAC+} which from Fig. 9.5 is $V_{buffer\text{-}CM}$, and hence the input of the unity gain buffer can have a different common mode voltage than the output. Thus in the MDAC topology of this chapter common mode voltage shifts between input and output in the unity gain amplifier do not affect the common mode voltages in subsequent pipeline stages.

To minimize the body effect an NMOS source follower requires a Deep N-Well layer (DNW). It is noted however that in most digital processes used by industry, the DNW layer is readily available and thus there is no additional fabrication cost in using the DNW layer.

The parasitic input capacitance of the source follower is determined by the capacitance from gate to drain (C_{gd}) and gate to source (C_{gs}) of transistor M1 as shown in Fig. 9.15.

From [10] it is noted that C_{gs} is the larger of the two input parasitic capacitors. A major advantage of the source follower topology is since the circuit has a unity gain between gate and source, the gate and source move approximately together, and thus the effect of the input capacitance C_{gs} is significantly reduced, leaving the input capacitance of the unity gain buffer to be dominated by the relatively small

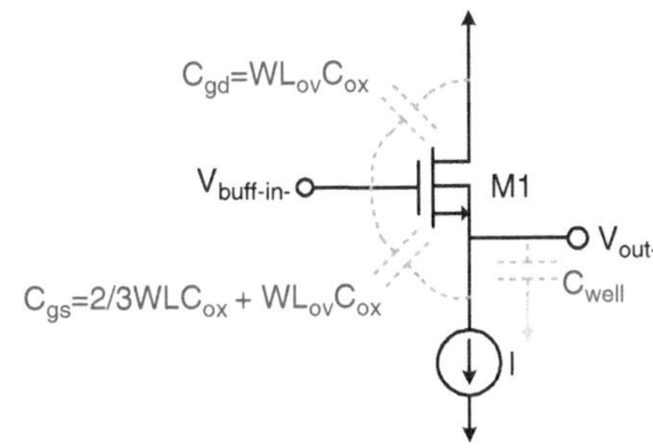

Fig. 9.15 Parasitic capacitance in NMOS source follower

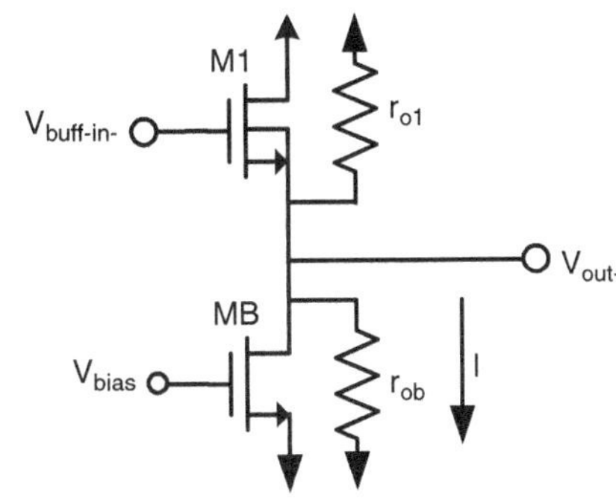

Fig. 9.16 NMOS source follower with output resistances labeled

C_{gd} [23]. Thus a source follower topology has small input parasitic capacitance, which hence enables a larger stage gain in the MDAC topology of this chapter.

Of note, the parasitic capacitance due to the DNW, C_{well}, slightly reduces the bandwidth, however as the output is discrete time, the non-linear nature of the well capacitance does not have a significant impact at the 10-bit level given a sufficient settling time.

9.5.1 *Linearity of Source Follower in a Sampled System*

The linearity of the source follower is limited primarily by the variation of the gain, and output impedance respectively with output signal swing. Since the source follower input is discrete time, non-linear settling times which plague source followers for sinusoidal inputs are not an issue for discrete time inputs assuming a sufficient settling time. If the body effect is eliminated by connecting the source and body together, the gain of a source follower as shown in Fig. 9.16 is given approximately by [23]:

$$A = \frac{g_{m1} r_{o1} || r_{ob}}{1 + g_{m1} r_{o1} || r_{ob}}, \tag{9.8}$$

where r_{o1} and r_{ob} are the output small signal resistances of M1 and MB respectively. Since $g_m \propto \sqrt{I}$ and $r_o \propto 1/I$, the gain of the source follower is a function of $I^{-0.5}$, and thus changes in the bias current 'I' with variations in the output voltage V_{out-} result in small changes in the gain of the source follower with output signal swing, hence harmonic distortion. The bias current 'I' is set predominantly by the bias voltage V_{bias}, however due to short channel effects, variations in the drain source voltage of MB also changes the bias current 'I'. Thus the signal swing at the output V_{out-} modulates the bias current 'I', which in turn results in distortion. To minimize distortion in the source follower several techniques can be used (e.g. [70]), however a simple solution is to use a large length for transistor MB, which trades a slightly larger load capacitance with lower design complexity. In this work harmonic distortion in the source follower was minimized by using a large length (L = 0.3 μm) for MB. An added advantage of using a large length for the bias transistor is that the transconductance of MB is reduced, and as will be seen in Section 9.6, results in lower noise in the source follower.

Thus in the MDAC topology of this chapter it is possible to achieve a linear stage-gain using a source follower, and hence the linear stage-gain calibration techniques as described in Section 5.2.2 can be used to achieve a high resolution in the ADC. Unlike [74], complicated non-linear calibration is not required in the approach of this chapter, even though opamps are not used.

9.5.2 *Signal Swing of Source Follower*

To maintain a linear output, the source follower is required to keep MB in the active region. Thus the minimum output voltage is given by the overdrive voltage of MB, i.e. $V_{out-} > V_{bias} - V_t$ (V_t is the threshold voltage). The maximum voltage is given by the fact that the source voltage of M1 in Fig. 9.17 is always V_t plus V_{ov1} below the gate (V_{ov1} is the overdrive voltage of M1). Thus if all voltages are limited to be no higher than V_{DD}, the maximum output voltage is V_{DD}-V_t-V_{ov} as noted in Fig. 9.17.

It is noted however, that since the source of M1 tracks the gate of M1 (i.e. approximately constant V_{gs}), in theory a voltage higher than V_{DD} could be used at the gate of the source follower without having to worry about reliability issues from oxide break down which occur when $V_{gs} > V_{DD}$. Thus assuming care is taken at

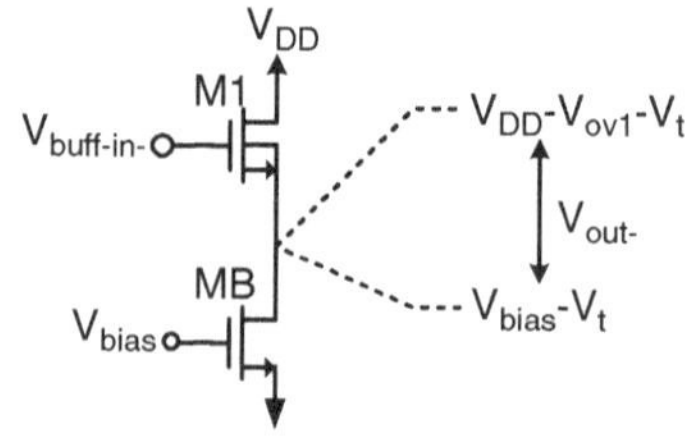

Fig. 9.17 NMOS source follower signal swing

startup and at all other voltage nodes in the MDAC, it is conceivable that a source follower could be used and a large signal swing realized.

Alternatively, it is noted that in industry dual supply voltages are commonly used, where the analog portions of an ADC are designed with a large supply voltages and thick oxide devices, whereas the digital portions are designed with thin oxide low voltage devices. Thus in a deep submicron process (e.g. 45 nm) where for example it is desired to keep all voltages below V_{DD}, a source follower could still be used as the unity gain amplifier so long as thick oxide devices and a second higher supply voltage are used for the source follower. Node voltages higher than V_{DD} and/or dual supply voltages were not used in this work; however an investigation of large signal swing (thus lower power) with the source follower is deferred as future research.

It is noted that if a low V_t NMOS (i.e. native device) was used in the source-follower, a significantly larger signal swing could be achieved, enabling smaller sampling capacitors and thus lower power consumption.

9.6 Noise Analysis of Capacitive Charge Pump Based MDAC

The input referred noise of the MDAC shown in Fig. 9.5 can be evaluated by referring all noise sources to one of the single-ended inputs of the ADC (e.g. V_{in+}). During Φ_1, the noise sampled by the capacitors is as shown in Fig. 9.18, and given by:

$$\sigma^2_{noise-\Phi 1} = \frac{kT}{C_1 + C_4}. \tag{9.9}$$

As shown in Fig. 9.19, during Φ_2 the noise is determined by the noise from the switches with an 'on' resistance of R_{on}, where the total noise that appears at V_{out-} is shaped by the bandwidth of the unity gain buffer which has a unity gain frequency of $\omega_{ta} = g_{m1}/C_L$.

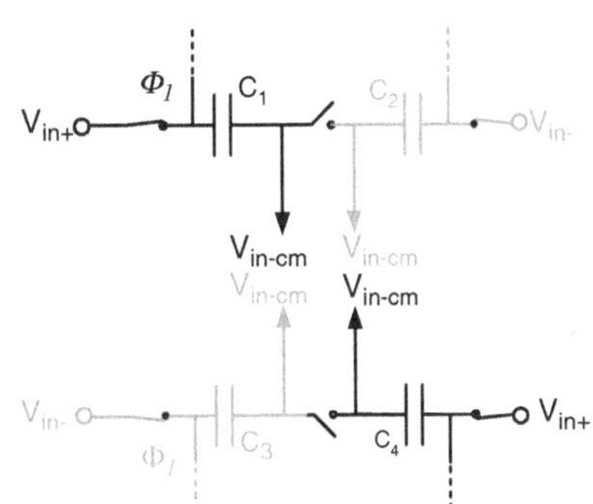

Fig. 9.18 Signal path of V_{in+} during Φ_1

Assuming the parasitic capacitances are only a small fraction of the sampling capacitors C_1, C_2, the noise during Φ_2 referred to the input V_{in+} can be found to be:

$$\sigma^2_{noise-\Phi 2} = \frac{1}{A^2}\left[2g_{m1}R_{on}\frac{kT}{C_L} + \sigma^2_{noise-buff}\right], \tag{9.10}$$

where A is the gain of the pipeline stage. As $V_{buff\text{-}in-}$ is effectively a floating node during Φ_2, capacitors C_1 and C_2 have no effect on the noise during Φ_2 under the assumption that the unity gain buffer is the block which determines the noise bandwidth during Φ_2. This is a fair assumption as R_{on} is typically small and the parasitic capacitors $C_{p\text{-}sw}$ and $C_{p\text{-}buff}$ are typically only a fraction of the sampling capacitors C_1, C_2.

When a source follower as shown in Fig. 9.20 is used for the unity gain buffer, the total noise of the buffer [23] can be found to be:

$$\sigma^2_{noise-buff} = \frac{4}{3}\frac{kT}{C_L}\left(1 + \frac{g_{m2}}{g_{m1}}\right) \tag{9.11}$$

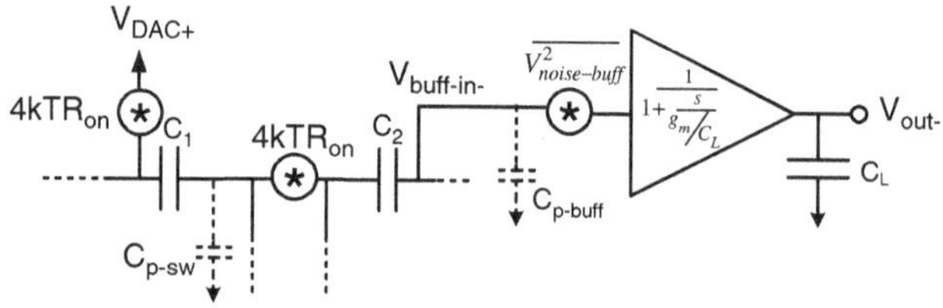

Fig. 9.19 MDAC configuration during Φ_2

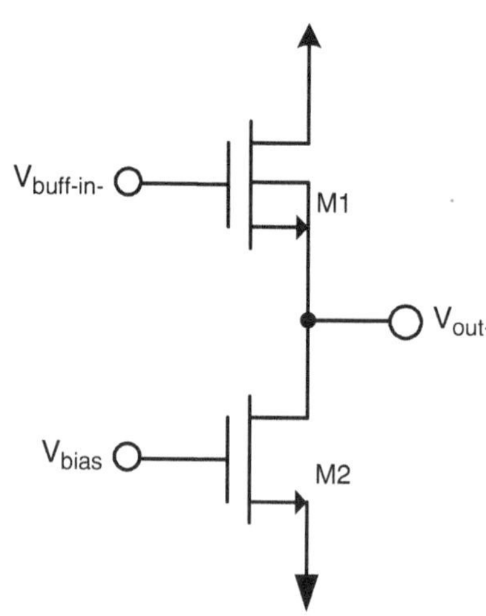

Fig. 9.20 NMOS source follower

Thus the noise in Φ_2 can be written as:

$$\sigma^2_{noise-\Phi 2} = \frac{1}{A^2}\left[2g_{m1}R_{on}\frac{kT}{C_L} + \frac{4}{3}\frac{kT}{C_L}\left(1+\frac{g_{m2}}{g_{m1}}\right)\right] \tag{9.12}$$

Hence the total input referred noise of the pipeline stage is given by the sum of equations (9.9) and (9.12):

$$\begin{aligned}\sigma^2_{noise-input} &= \sigma^2_{noise-\Phi 1} + \sigma^2_{noise-\Phi 2}\\ &= \frac{kT}{C_1+C_4} + \frac{1}{A^2}\left[2g_{m1}R_{on}\frac{kT}{C_L} + \frac{4}{3}\frac{kT}{C_L}\left(1+\frac{g_{m2}}{g_{m1}}\right)\right]\end{aligned} \tag{9.13}$$

Figure 9.21 shows a Spice simulation of the power spectral density of the noise output of the circuit shown in Fig. 9.19 during Φ_2, when $C_L = 1.06$ pF, $g_{m1} = 2.2$ mA/V, $g_{m2} = 0.84$ mA/V, and $T = 300$ K.

From equation 9.12, when the thermal noise from the switches is assumed small, the output referred noise (i.e. before dividing by the gain squared) is calculated to be:

$$\begin{aligned}\sigma^2_{noise-\Phi 2-output} &\approx \frac{4}{3}\frac{kT}{C_L}\left(1+\frac{g_{m2}}{g_{m1}}\right)\\ &= \frac{4}{3}\frac{(1.38e-23)(300k)}{1.06pF}\left(1+\frac{0.84}{2.2}\right)\\ &= 7.2\times 10^{-9}\mathrm{V}^2,\end{aligned}$$

which agrees very well with the simulated results, thus verifying the derived noise analysis.

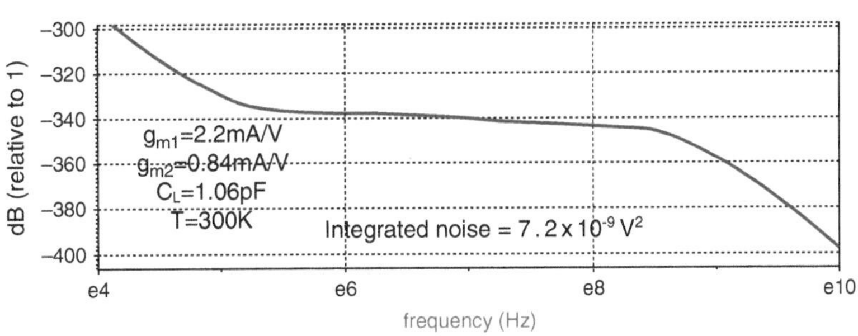

Fig. 9.21 Power spectral density of noise at V_{out-} during Φ_2

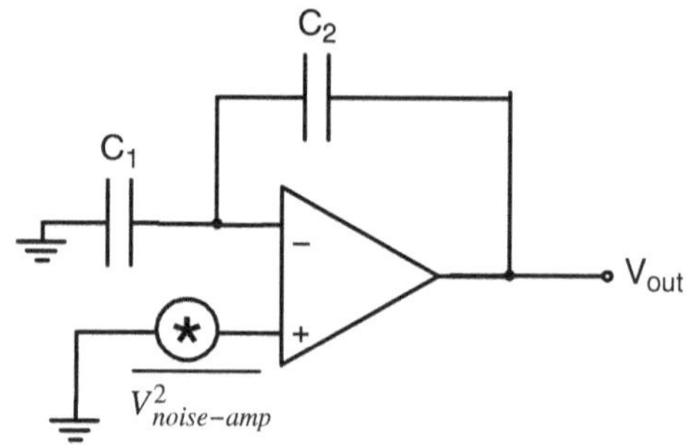

Fig. 9.22 Amplifier noise during Φ_2 in traditional MDAC

In general R_{on} can be made small, and $g_{m2} < g_{m1}$, thus the total input referred noise can be approximated by:

$$\sigma^2_{noise-input} \approx \frac{kT}{C_1 + C_4} + \frac{1}{A^2}\left(\frac{4}{3}\frac{kT}{C_L}\right) \tag{9.14}$$

In a traditional 1.5-bit based MDAC topology as shown in Fig. 8.1, during Φ_1, the noise is also given by $kT/(C_1 + C_2)$. To find the noise during Φ_2, consider Fig. 9.22 which shows the basic configuration of the traditional MDAC topology during Φ_2.

If it is reasonably assumed that the switch bandwidth is very large, the noise during Φ_2 will be dominated by the opamp. From Fig. 9.22, during Φ_2 the DC gain of the noise is given by [95]:

$$A_{noise-\Phi 2} = 1 + \frac{C_1}{C_2} = 2 \tag{9.15}$$

(assuming $C_1 = C_2$). However the signal gain is also 2, thus the thermal noise of the opamp directly contributes to the input referred noise of the pipeline stage. If it is assumed the noise of the opamp is dominated by the input differential pair, the opamp thermal noise referred to the input of the MDAC can be found to be [96]:

$$\sigma^2_{noise-buff} \approx \frac{4}{3}\frac{kT}{C_{L2}} \tag{9.16}$$

where C_{L2} is the total load capacitance during Φ_2. Hence the total input referred noise for the traditional MDAC topology is given by:

$$\sigma^2_{noise-traditional} \approx \frac{kT}{C_1 + C_2} + \frac{4}{3}\frac{kT}{C_{L2}} \tag{9.17}$$

Comparing equations (9.14) and (9.17) it is clear that the MDAC topology of this work has a significant advantage in that since the unity gain buffer is connected to the output of the passive gain stage, the noise of the amplifier is reduced by A^2 when referred to the input. Thus the MDAC can be designed to achieve the same speed as an opamp based MDAC with smaller capacitors and thus lower power than the traditional MDAC.

9.7 Calibration of Pipeline Stages

In Section 3.4 it was noted that gain errors in the MDAC resulted in missing codes which increase harmonic distortion in the ADC's output. From equation (9.6) it is clear that the gain of the 1.5-bit stage is a function of parasitic capacitance as well as the gain of the buffer – values which cannot be accurately estimated before fabrication. As discussed in Section 5.2 however, calibration can be used to measure and correct the non-ideality. In this work foreground calibration was used to measure and correct the error of each stage. From equation (9.6) however, the gain of the pipeline stage in this chapter is a function of the unity gain amplifier's gain, which for a source follower is a function of temperature which varies with time. Thus in a practical implementation for use in industry it would be more desirable to use a background calibration scheme. Implementing a background calibration scheme however as discussed in Chapter 6 is non-trivial, thus in the interest of reducing the complexities of an initial prototype, foreground calibration [97] – which is much simpler to implement, was used. It is noted that the low power MDAC topology of this chapter is compatible with most background calibrations schemes, including the split-ADC approach discussed in Chapter 4. Thus it is conceivable that in an industrial implementation background calibration could be used in the ADC.

9.7.1 Foreground Calibration in Detail

Consider the ADC topology of Fig. 9.23, which shows a 1.5-bit first pipeline stage followed by an ideal backend Flash ADC.

If there is a gain error in the first pipeline stage, the output of the ADC is as shown in Fig. 9.24.

Thus the objective of the calibration scheme is to estimate the number of missing codes, ε.

Consider the residue transfer curve of a 1.5-bit stage as shown in Fig. 9.23. If the input to the pipeline stage is zero, the DAC voltage can be either 0, $+V_{ref}$, or $-V_{ref}$, and assuming no gain errors in the pipeline stage, the output of the 1.5-bit pipeline stage will not saturate the output due to the 0.5-bit redundancy of the pipeline stage. Thus in an ideal 1.5-bit pipeline stage with zero input, the output of the ADC will be constant regardless of the DAC voltage (assuming the MSB bits reflect the value of the DAC voltage).

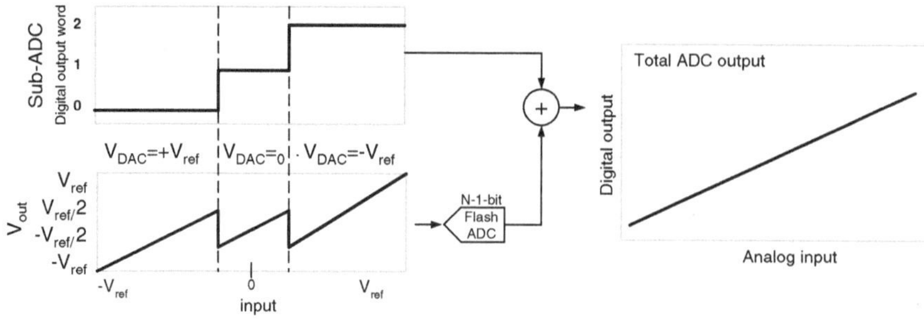

Fig. 9.23 Ideal 1.5-bit first pipeline stage

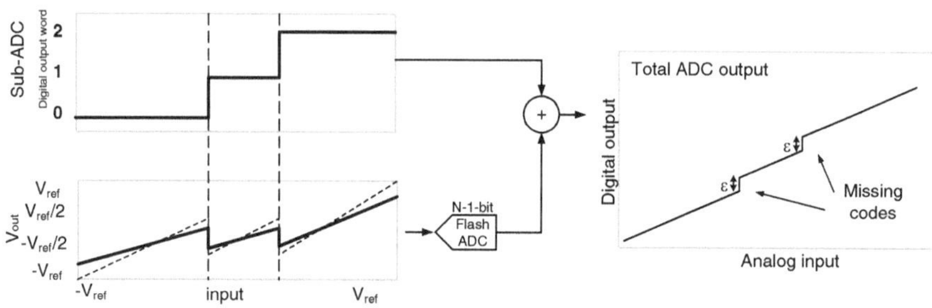

Fig. 9.24 1.5-Bit pipeline stage with gain error

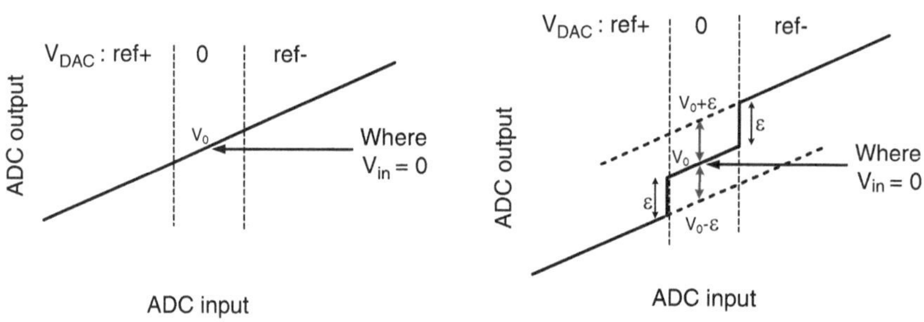

Fig. 9.25 Measure of missing codes when pipeline stage input (V_{in}) is zero – left is ideal, right is with errors

However if there is a gain error in the pipeline stage (such that the gain is less than the ideal value), the ADC will output different values when the DAC voltage is tied to $+V_{ref}$, 0, and $-V_{ref}$ respectively with the first pipeline stage set to have a zero input. Figure 9.25 illustrates the ADC output when V_{in} is near zero without and with errors.

Thus the missing codes produced by a non-ideal stage gain can be corrected in the foreground by shorting the input of the pipeline stage under calibration to zero,

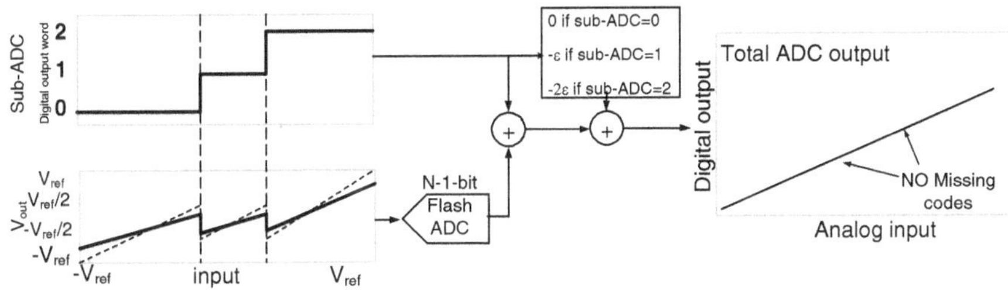

Fig. 9.26 Illustration of correction scheme

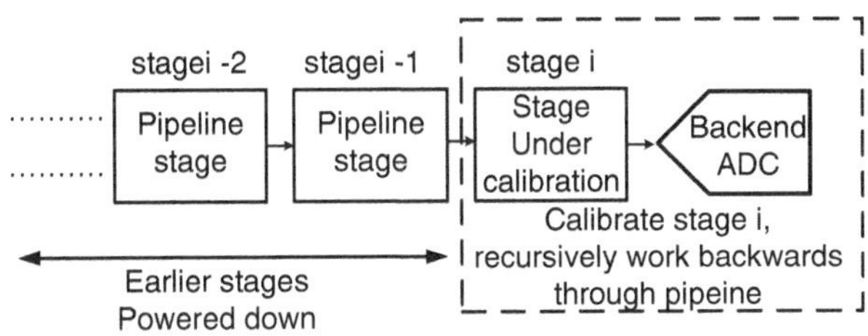

Fig. 9.27 Multistage foreground calibration

and separately measuring the output of the ADC when the DAC voltage of the stage under calibration is connected to $+V_{ref}$, 0, $-V_{ref}$ respectively. By averaging out each value for a few clock cycles to suppress thermal noise, an accurate estimate of the error ε can be found by subtracting the average ADC output when $V_{DAC} = V_{ref-}$ from the average output when $V_{DAC} = 0$, and/or similarly by subtracting the average ADC output when $V_{DAC} = 0$ from the average output when $V_{DAC} = V_{ref+}$. The gain error is subsequently corrected by shifting the digital output by the amount of the missing codes as shown in Fig. 9.26.

All pipeline stages are calibrated in this work using a foreground approach. Multiple stages are calibrated at startup by recursively using the method discussed this section initially on the last pipeline stage (while powering off all previous stages), then the second last, then the third last, etc., eventually calibrating the entire pipeline ADC as shown in Fig. 9.27 (e.g. [98]).

9.8 Theoretical Power Savings

In this section the theoretical power consumption of the MDAC topology of this chapter is compared against the power consumption of a traditional MDAC. The goal of the analysis is to examine under what circumstances the topology of this chapter offers a benefit of substantial power savings over the traditional MDAC architecture.

From Section 3.5 it was noted that the unity gain frequency of an opamp in closed-loop is given by: $\omega_t = \frac{\omega_{ta}}{1/\beta}$, where β is the feedback factor. However in the MDAC topology of this chapter since the only active circuit (source follower in this work) requires only a unity gain, β = 1. Hence on bandwidth considerations alone the topology of this chapter improves the settling time of the MDAC by feedback factor of the traditional MDAC topology. To make a fair comparison between the passive gain and traditional MDAC topologies, the traditional MDAC topology should be designed with the same closed-loop gain as is achieved in the MDAC of this chapter, noting that since the MDAC of this chapter achieves a gain of less than 2x (due to parasitic capacitors as noted in Section 9.4), the traditional MDAC can trade more bandwidth for less gain to match the MDAC topology of this chapter. Thus if the stage gain of the MDAC of this chapter is given by 'A', at best β = 1/A for the traditional MDAC, hence considering only bandwidth the power savings of the approach of this work (η) is given by:

$$\eta = \frac{\omega_{t-proposed}}{\omega_{t-traditional\_MDAC}} = \frac{\omega_{ta}}{\omega_{ta}/\beta} = \beta = \frac{1}{A}. \tag{9.18}$$

It was noted in Section 9.6 that one of the benefits of the approach of this work was: since the active circuit comes after the gain stage, the noise of the active circuit is reduced by the gain when referred to the input. From equations (9.14) and (9.17), the ratio (ζ) of input referred noise between the MDAC of this chapter and the traditional MDAC of Fig. 8.1 is:

$$\varsigma = \frac{\sigma^2_{T-proposedMDAC}}{\sigma^2_{noise-traditional}} = \frac{\sigma^2_{T\_\Phi1} + \sigma^2_{T\_\Phi2}/A^2}{\sigma^2_{T\_\Phi1} + \sigma^2_{T\_\Phi2}} \tag{9.19}$$

To simplify the analysis assume $\sigma^2_{T\_\Phi1} = \sigma^2_{T\_\Phi2}$. Thus:

$$\varsigma \approx 0.5 + \frac{1}{2A^2} \tag{9.20}$$

Hence to achieve the same input referred noise floor in both the passive gain and traditional MDAC topologies the MDAC of this chapter can be designed with 1/ ζ smaller capacitors (assuming the same input signal swing). Since [10]:

$$\omega_{ta} = \frac{g_m}{C_{load}} = \frac{2I_D}{C_{load}V_{eff}}, \tag{9.21}$$

it can thus be inferred that since the capacitors of MDAC of this chapter are reduced from the traditional MDAC by 1/ζ, the power of the MDAC of this chapter can also be reduced by 1/ζ while preserving the same speed. Thus the total factional reduction in power of the MDAC of this chapter versus a traditional MDAC (defined as α) is given by:

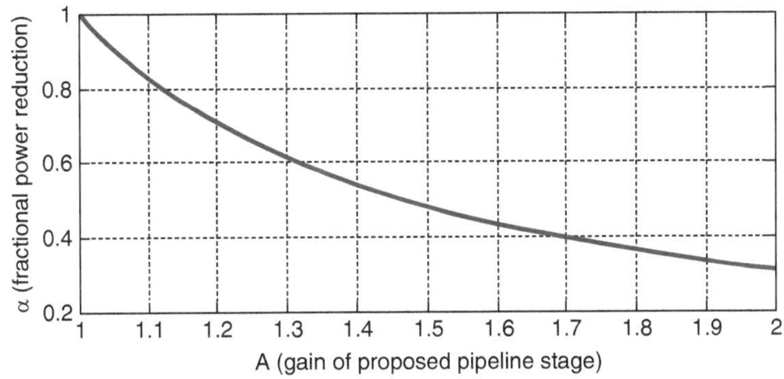

Fig. 9.28 Fractional reduction of power in MDAC of this chapter versus traditional MDAC

$$\alpha = \eta\varsigma = \frac{1}{A}\left(0.5 + \frac{1}{2A^2}\right) \tag{9.22}$$

Figure 9.28 plots α versus the gain 'A' of the pipeline stage in this chapter.

From Fig. 9.28 it is seen that the efficiency of power reduction in the MDAC of this chapter is maximized when the stage gain of the MDAC is maximized. It is noted that the efficiency plot of Fig. 9.28 ignores the effect of parasitic capacitors in the opamp (which further reduce β), as well as the fact that in practice the noise of an opamp is usually larger than that of a source follower. These approximations were made as in practice the actual value of β and noise from the opamp varies from one implementation to another, thus the best case β and opamp noise were used in the analysis to show the minimum amount of power reduction possible using the approach of this chapter. Hence the power reduction of the MDAC of this chapter is even larger than that shown in Fig. 9.28.

The power consumption of the additional digital calibration circuitry required in the ADC of this chapter has been ignored in the analysis, as it is noted that strictly speaking foreground calibration adds only a small amount of additional power since it only powers on periodically, and the power of digital adders can be made small. It is noted that in general digital background calibration of linear errors in deep sub micron processes typically only add 10–20% extra power, where the specific amount of additional power varies depending on which background calibration technique is used, and how frequently the background calibration is powered on.

9.9 Design Specifications

From Section 9.8 it was noted that the MDAC topology of this chapter showed the greatest improvement in power consumption over the typical MDAC topology when the stage gain was maximized. To maximize operating speed, all switches need to be sized large to minimize the settling time, and the unity gain buffer needs to be large to maximize bandwidth. However, from equation (9.6) it is seen that the larger the parasitic capacitors (thus larger the switches, and larger the unity gain amplifier) the smaller the stage gain. Hence there is a tradeoff in this work of higher speed with lower power efficiency. The design specifications of a 10-bit 50 MS/s ADC were ultimately selected as Spice simulations showed at 50 MS/s stage gains on the order of $\sim\times 1.8$ could be achieved using the MDAC topology of this chapter. Furthermore as this book already deals with the design of two power efficient opamp based 10-bit 50 MS/s ADCs in Chapters 7 and 8, a fair comparison can be made between the topology of this chapter and the ADC discussed in the earlier chapters. Although a specific application was not targeted with this work as it was a proof-of-concept prototype, a quick survey of 10-bit ADC with sampling rates on the order of 50 MS/s show a variety of potential applications from medical imaging to digital communications.

9.10 Circuit Design

In this section the circuits used in the design of a prototype of a 10-bit 50 MS/s ADC in 1.8 V, 0.18 μm CMOS which used the power efficient MDAC topology of this chapter are described.

9.10.1 ADC Top Level Topology

Figure 9.29 illustrates the top level topology of the ADC designed using the power reduction techniques discussed in this chapter.

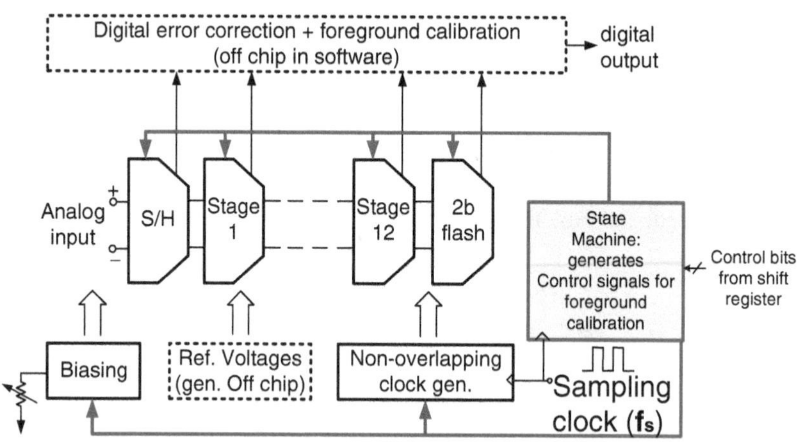

Fig. 9.29 Top-level topology of ADC used in the work of this chapter

Simulation results showed each pipeline stage to have a stage gain of $\sim\times 1.8$, i.e. $\log_2(1.8) = 0.85$ true bits resolved per stage (where 'true bits' denotes the number of bits resolved which reduce the quantization noise floor, i.e. ignoring the redundant bits). Thus with 12 total stages followed by a 2-bit Flash ADC, approximately 12 true bits are resolved ignoring the thermal noise. As ADC power is dominated by thermal noise considerations, the thermal noise floor was designed to be at the ~10-bit level at the input of the ADC. To minimize power, the first three pipeline stages were scaled approximately by their respective stage gains [16].

9.10.2 Front-End Sample-and-Hold

To simplify the design of the ADC and focus all design efforts on the novel power reduction techniques, no attempt was made to eliminate the front end S/H.

The front-end S/H was also realized in this work without opamp-based capacitive feedback as shown in Fig. 9.30, where rather than using a passive voltage gain technique as used in the MDAC, the sampled input is simply buffered by a source follower [99].

Switch S6 is included so that during Φ_1 the source follower can be powered off, and thus save additional power. $V_{in\text{-}cm\text{-}SH}$ was set as 0.4 V and $V_{buffer\text{-}CM}$ was set as 1.4 V, where each voltage was generated off-chip.

Although not done in this work it is possible that a passive gain of two could be used in the S/H as shown in Fig. 9.31 so as to relax the signal swing requirements and thus linearity requirements of the input sampling switches SW_INP, SW_INN.

9.10.3 MDAC and Unity Gain Amplifier

Figure 9.32 shows the full circuit topology of the first stage MDAC used in this work. Subsequent pipeline stages were identical where the first three pipeline stages were scaled by the gain of the pipeline stage. An NMOS source follower using

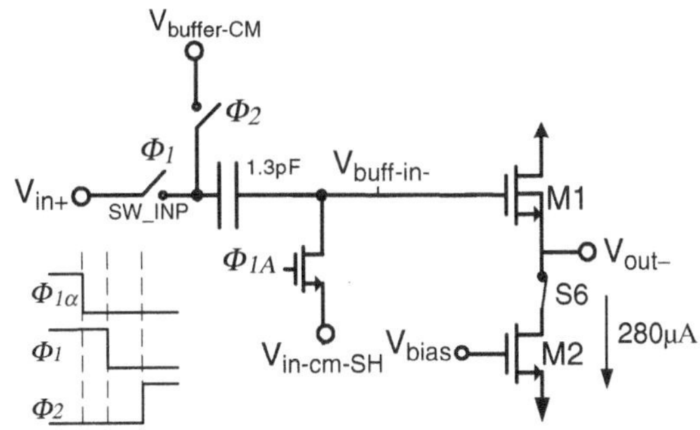

Fig. 9.30 Front-end sample-and-hold using unity gain buffer (shown single-ended, implemented pseudo-differentially in practice)

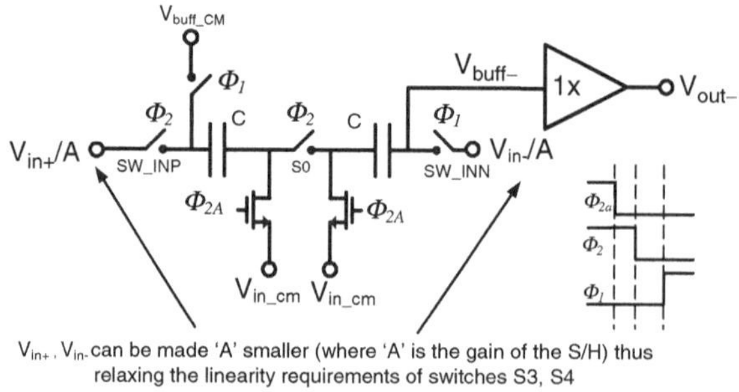

Fig. 9.31 Modified S/H which has a gain of 'A'

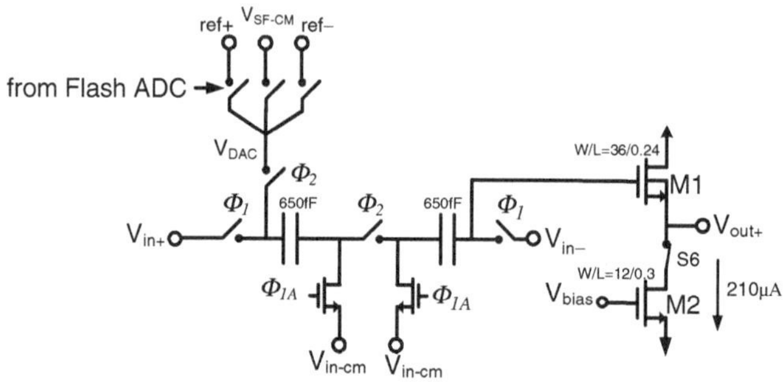

Fig. 9.32 First stage MDAC circuit

a DNW layer was used to implement the unity gain amplifier. To further reduce power consumption a switch S6 was added to power off the unity gain amplifier during Φ_1 as it is not required to operate during Φ_1.

As noted in Section 9.5.1, to maximize linearity and minimize noise, the length of M2 was 0.3 μm. To maintain linearity beyond the 10-bit level, the output signal swing of the source follower was designed to be 0.5 V p-p single-ended (i.e. 1 V p-p when considering the differential output). It is noted that the ADC in this chapter has a signal swing smaller than the differential 1.6 V p-p of the ADC discussed in chapter 5. This is because simulations show a signal swing larger than 1 V p-p results in increased nonlinearity from the buffer such that more than 10-bits linearity cannot be achieved. Thus a clear tradeoff in using a simple source follower buffer is reduced signal swing, thus increased capacitor sizes in the MDAC, hence increased power consumption. However as will be seen in Section 9.12, even with a ~40% smaller signal swing, the topology of this chapter is so power efficient that a significant power reduction still occurs when comparing this work versus that of Chapters 7 and 8.

The sizes of all the switches were optimized based on simulation results, where all the switches were large enough to achieve the desired sampling rate of 50 MS/s,

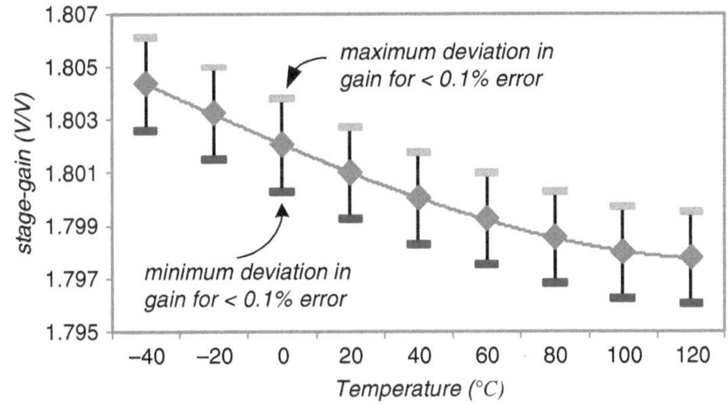

Fig. 9.33 Stage-gain variation with temperature (Based on simulation)

but small enough to minimize distortion and parasitic capacitances on critical nodes.

The values of the reference voltages for the MDAC, which were generated off-chip, were: $V_{ref+} = 1.55$ V, $V_{ref-} = 1.05$ V, $V_{cm\text{-}buff} = 1.3$ V, $V_{in\text{-}cm} = 0.5$ V.

Figure 9.33 shows the variation of the stage-gain of Fig. 9.32 over temperature based on simulation results. From Fig. 9.33 it is clear that while the gain does vary over temperature the variation is below the 10-bit level for a wide variation in temperature, thus frequent recalibrations due to temperature fluctuations may not be required. If a higher resolution ADC were targeted and/or the ADC used in a system which had drastic temperature variations, a background calibration scheme could alternatively be used to ensure temperature induced gain fluctuations were always accounted for.

9.10.4 Sub-ADC

The sub-ADC was designed using the same dynamic comparators used in the ADC of Chapters 7 and 8 as shown in Fig. 9.34. Dynamic comparators have the advantage of low power consumption, but at the cost of increased offset. This however is a favorable tradeoff as the 1.5-bit topology has a large amount of redundancy to trade with comparator offset [15].

As the sub-ADC connects to the output of a source follower which has a common-mode voltage near V_{SS}, the comparators required different reference voltages than those supplied to the DACs of each pipeline stage. Since the pipeline topology can tolerate offsets from the comparator, different values of the reference voltages can be provided to the sub-ADC provided they do not incur an effective offset in each comparator larger than $V_{ref}/4$. Furthermore by separating the reference voltages in the comparators from the DAC, the amount of switching noise on

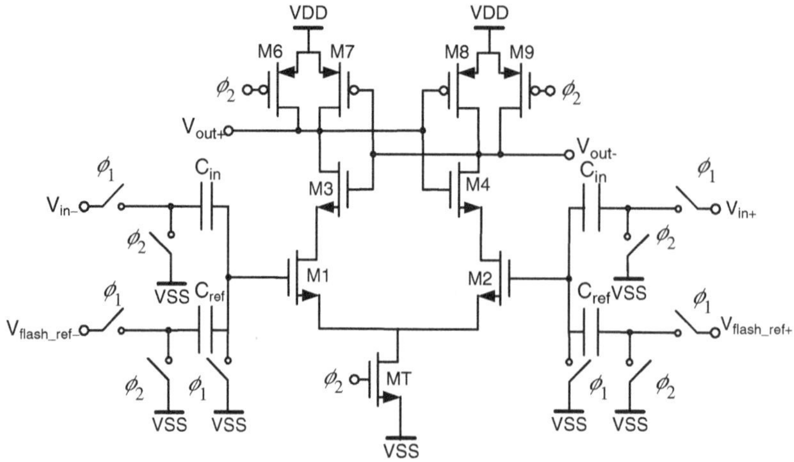

Fig. 9.34 Dynamic comparator used in flash sub-ADC

the reference voltages is reduced. The comparator's reference voltages which were generated off-chip were: $V_{flash\_ref+} = 0.75$ V and $V_{flash\_ref-} = 0.25$ V.

9.10.5 Digital State Machine

A digital state machine was used to generate the control signals for each pipeline stage during foreground calibration. The state machine was only powered on during foreground calibration and powered completely off subsequently. The state machine was ultimately programmed so that complete calibration was completed within 10^4 clock cycles.

9.10.6 Analog Test-Mux

To enhance the testability of the ADC each pipeline stage was equipped with three analog test muxes: two to enable viewing of the differential input to each pipeline stage, and a third test mux to observe the bias voltage of the source follower. The goal of the test mux was to be able to verify basic functionality without adding a significant additional capacitive load. As such, each analog mux was implemented as a transmission gate which was sized approximately the same as the transmission gates for the sampling switches. Thus basic functionality of the ADC can be verified by running the ADC at low sampling rates, avoiding otherwise non-trivial wafer probing.

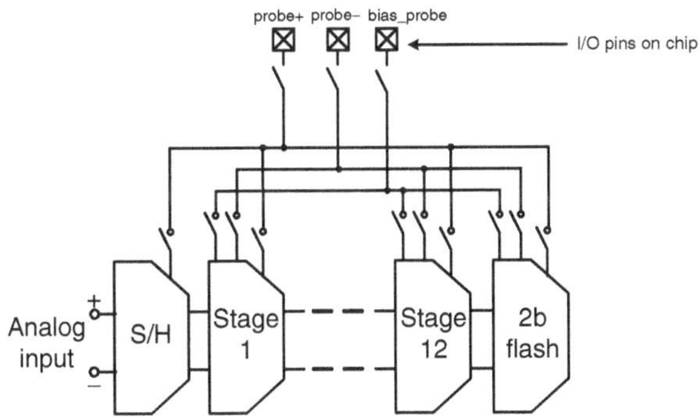

Fig. 9.35 Analog test mux configuration

The three analog test-muxes in each pipeline stage were all connected in parallel to three I/O pins in the chip as shown in Fig. 9.35. Control bits loaded via a shift register were used to select which pipeline stage connected to the I/O pins.

9.11 Testing

Sections 9.11.1 and 9.11.2 discuss the test setup and measured results of the prototype fabricated in a 1.8 V, 0.18 μm CMOS process.

9.11.1 PCB

A four layer FR4 dielectric PCB board with a minimum 6 mil trace was designed and constructed for the device under test as shown in Fig. 9.36. Separate Power planes were used to isolate the analog, digital, I/O, and board power supplies. A differential input was generated using a 1:1 turns ratio Minicircuits transformer matched to 50 Ω. Reference voltages were generated by passing the output of a resistive voltage divider through an opamp (LM7301) in a unity gain buffer configuration. To maintain constant supply voltages, all voltage supplies for each power plane were generated through regulators (LM337, LM1117), and heavily decoupled with capacitors. As the ADC utilized a constant current biasing scheme, an off-chip adjustable resistor was used as the master current source.

9.11.2 Test Setup

A test setup as shown in Fig. 9.37 was used. Sinusoidal inputs were generated using a HP 8664A function generator. Several Minicircuits filters were used to minimize

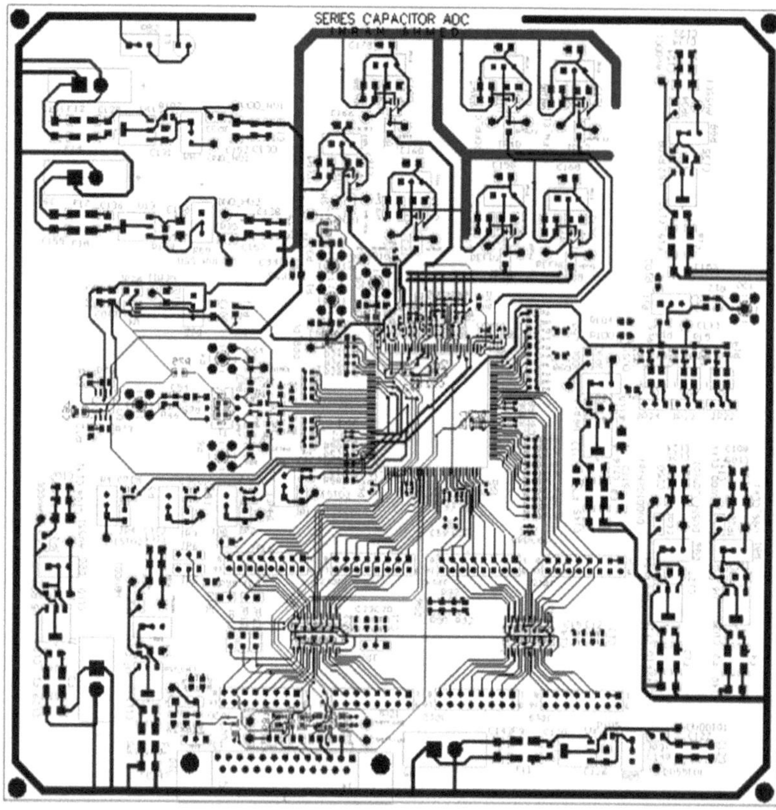

Fig. 9.36 PCB used to in test setup for ADC described in this chapter

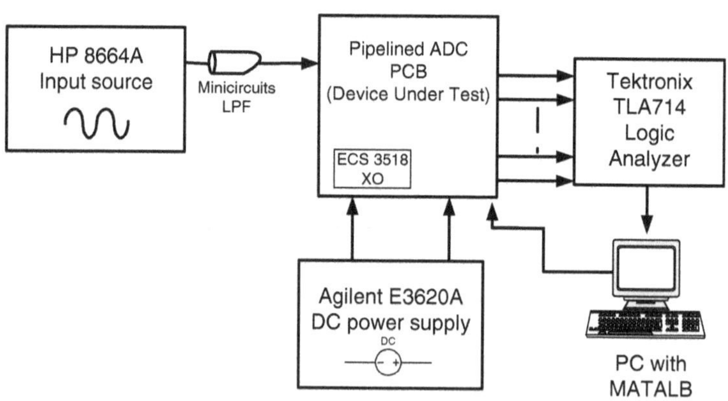

Fig. 9.37 Test setup of ADC

harmonic distortion from the function generator such that the sinusoidal input to the ADC had an SNDR of well over 62 dB for input frequencies larger than 21 MHz. A 50 MHz crystal oscillator (ECS3518-XO) on the PCB was used to generate the clock for the ADC. The serial shift register was loaded via a parallel port connection to a PC, where a Matlab script was executed to load the appropriate bits. The output bits of each pipeline stage were captured using a Tektronix TLA714 logic analyzer, capable of capturing 65,536 points at a time. An Agilent E3620A Dual output DC power supply was used to provide positive and negative voltages to the voltage regulators on the PCB. The 10-bit output word from the 10-bit ADC was determined via a Matlab script written to emulate the operation of a digital error correction circuit and foreground calibration.

9.12 Measured Results

A prototype of the ADC of this chapter was fabricated in a 1.8 V 0.18 μm CMOS process as shown in Fig. 9.38; the core area was 2.0 mm × 0.7 mm (1.4 mm^2). The fabricated IC was packaged in a 120-pin CQFP package, where 78 of the total 120 pins were used. A large number of pins were used to allow many pins to be used for DC power supplies and reference voltages, so as to reduce the impact ringing due to the parasitic inductance of bond wires.

As noted in Fig. 9.38 approximately a quarter of area in the chip was dedicated to test circuitry such as digital muxes which enabled each digital signal in each pipeline stage to be configured in different modes for testability. In a practical implementation much of the test circuitry can be removed hence allowing for a reduction in area consumption.

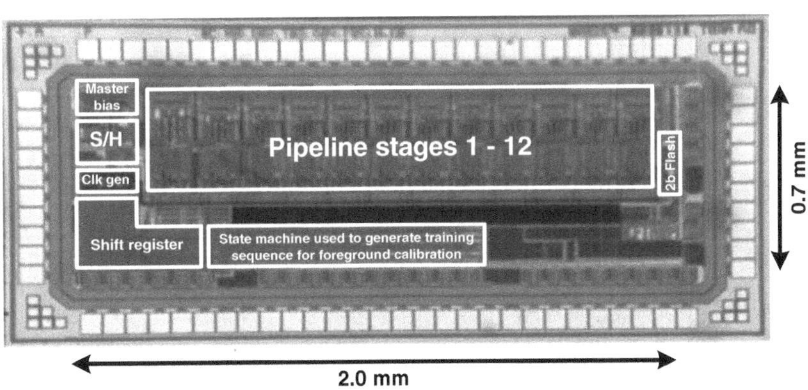

Fig. 9.38 Micrograph of low powered pipeline ADC

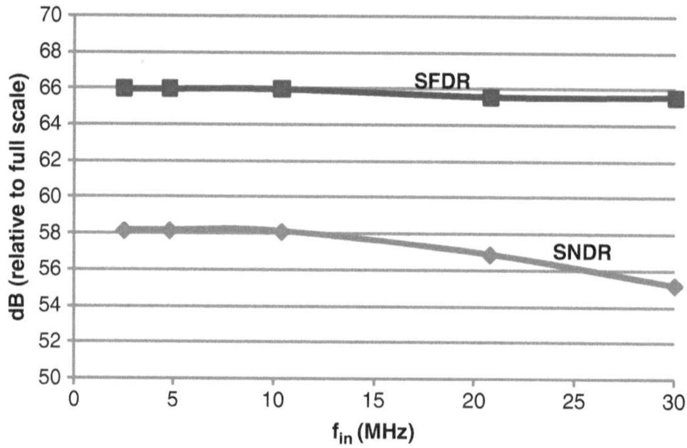

Fig. 9.39 SNDR/SFDR variation with input frequency, $f_s = 50MS/s$

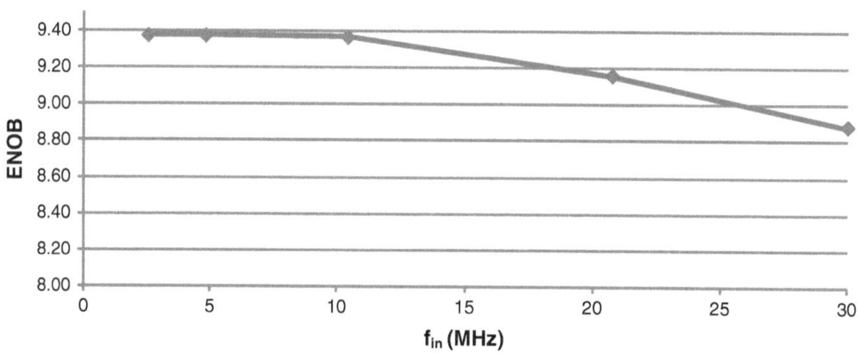

Fig. 9.40 ENOB variation with input frequency, $f_s = 50MS/s$

9.12.1 *Measured ADC SNDR Variation*

Figure 9.39 shows the variation of the SNDR and SFDR of the ADC at 50 MS/s for input frequencies between 2.4 and 30 MHz, where it is seen that the ADC is capable of achieving an SNDR/SFDR as high as 58.2/66 dB.

Figure 9.40 shows the variation of ENOB with input frequency where it is seen the ADC can achieve as high as 9.4-bits.

The ADC consumed only 3.9 mW for the active amplifiers and 6 mW for all the clocking circuits, resulting in a total power consumption of only 9.9 mW. Although the power of the reference voltages is not included, it is noted that the total average current demanded by ADC from the off-chip reference voltages was only 0.34 mA. The ADC of this work consumed less than half the power of the ADC discussed in Chapters 7 and 8 (9.9 versus 35 and 27 mW, respectively), even though the ADC of

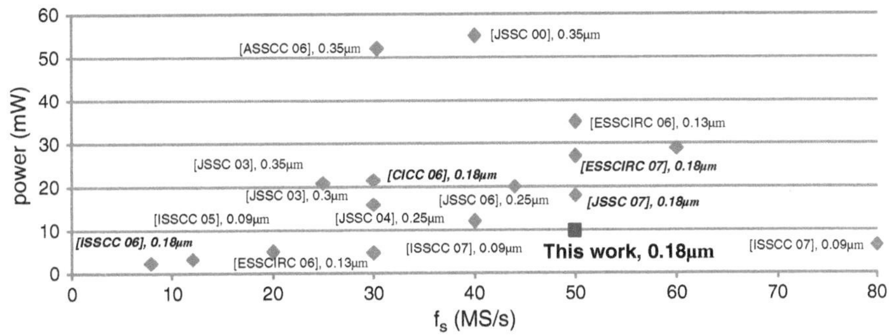

Fig. 9.41 Comparison of power of ADC of this work versus other 10-bit ADCs

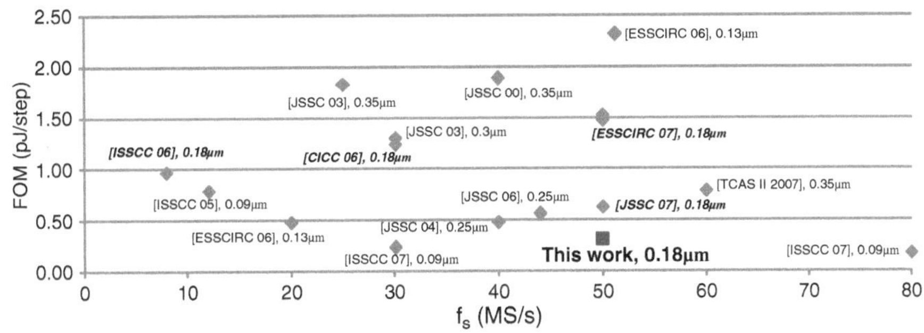

Fig. 9.42 Comparison of FOM of ADC of this work versus other 10-bit ADCs

this chapter had a smaller signal swing (1.0 versus 1.6 V p-p) and included a front-end sample-and-hold. It is conceivable if a larger signal swing was used and the front end sample-and-hold removed, the ADC discussed in this chapter could have a further reduction in power. The fact that the majority of power consumed by the ADC of this work is dynamic suggests that a large reduction in power could be achieved by lowering the digital/clocking supply voltage and/or migrating to a smaller technology.

Figure 9.41 compares the power of the ADC of this work versus other recently published 10-bit ADCs where it is seen that amongst 10-bit pipeline ADCs the architecture of this chapter has the lowest power consumption at 50 MS/s.

Figure 9.42 compares the figure of merit using equation (2.2) for 10-bit pipelined ADCs where it is seen that the ADC of this work has amongst the best published figure of merits for 10-bit ADCs in the 10–80 MS/s range. The few publications which have a slightly better figure of merit have the benefit of a faster technology (0.18 μm publications are shown in bold italics). Furthermore it is noted that amongst 0.18 μm ADCs, this work achieves the best figure-of-merit.

Figures 9.41 and 9.42 clearly illustrate the significant power savings afforded by the ADC topology of this work.

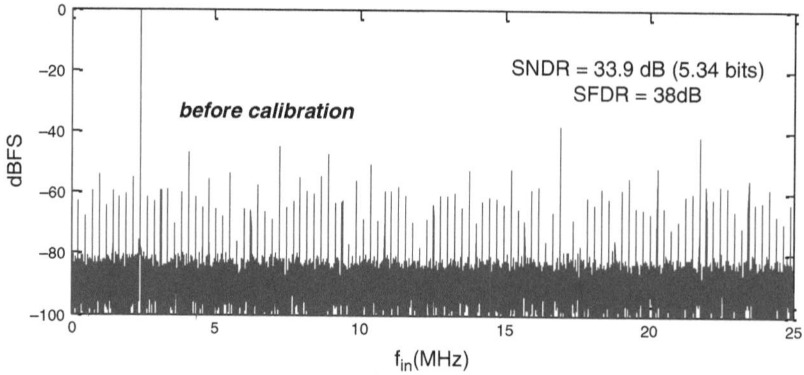

Fig. 9.43 FFT of ADC output before calibration, f_{in} = 2.4 MHz, f_s = 50 MS/s

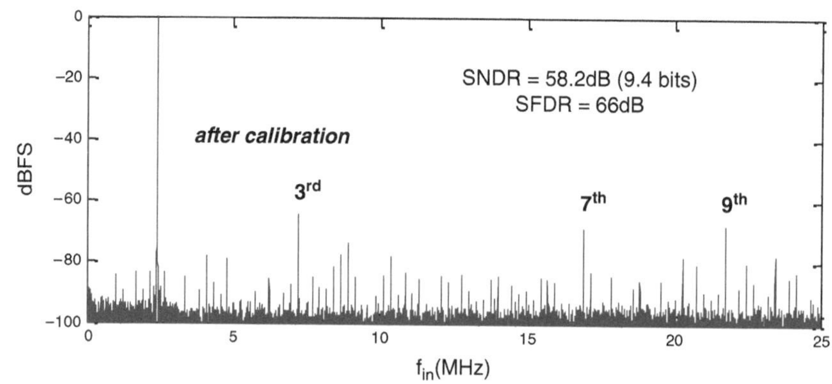

Fig. 9.44 FFT of ADC output after calibration, $f_{in} = 2.4$ MHz, $f_s = 50$ MS/s

9.12.2 ADC FFTs

Figures 9.43–9.46 show FFTs of the ADC output for input frequencies of 2.4 MHz and 20.7 MHz before and after foreground calibration for $f_s = 50$ MS/s.

From the FFTs it is clear that calibration is clearly heavily leveraged to achieve a significant improvement in ADC performance – almost 4-bits.

From the FFT plots after calibration, it can be seen that the even order distortion terms are heavily attenuated, verifying that the sampling topology shown in Fig. 9.5 achieves differential functionality. The degradation of ADC accuracy for higher sampling rates was attributed to distortion from the input sampling switch (switch SW_IN(P/N) in Fig. 9.30), and increased digital noise from the I/O buffers. The distortion due to the sampling switch could be reduced with only a small increase in

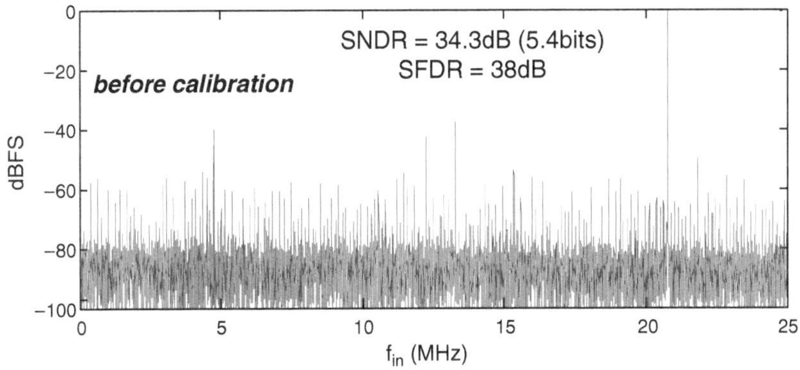

Fig. 9.45 FFT of ADC output before calibration, $f_{in} = 20.7$ MHz, $f_s = 50$ MS/s

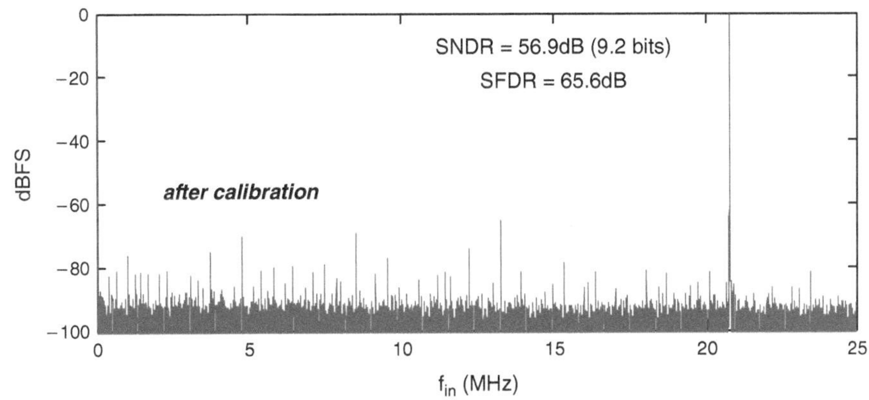

Fig. 9.46 FFT of ADC output after calibration, $f_{in} = 20.7$ MHz, $f_s = 50$ MS/s

power by either using larger transistors in the input switch and/or using bootstrapping techniques [77].

It is noted that when the bias currents of the ADC were varied by more than $\pm$ 20%, the ADC SNDR varied by less than 0.1 bits when using correction terms extracted when the ADC was biased with nominal bias currents. These results indicate that the gain of each pipeline stage was set primarily by capacitor ratios – which do not change over time. Hence in a practical implementation, the interval between foreground calibrations could be very large, allowing for a minimal impact on normal ADC operation. Drift of correction terms is also negligible, as measured results from the ADC separated by 1 week show insignificant (less than 0.1 bits) difference in ADC ENOB when the same correction terms are used for both measurements.

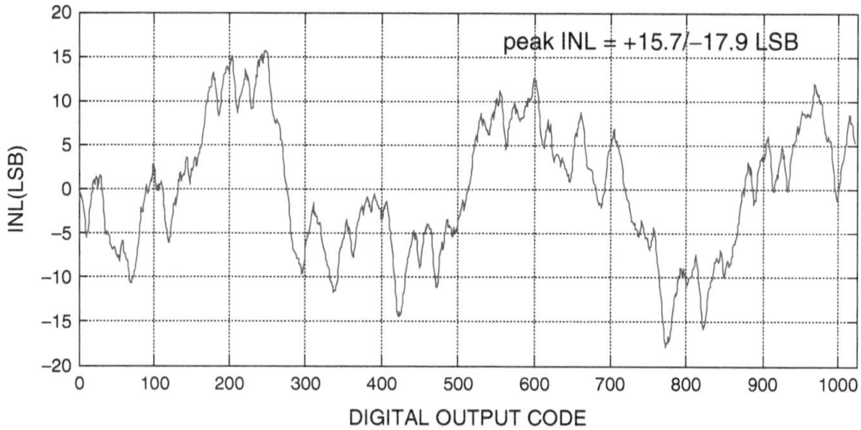

Fig. 9.47 INL before calibration (LSB @ 10-bit level)

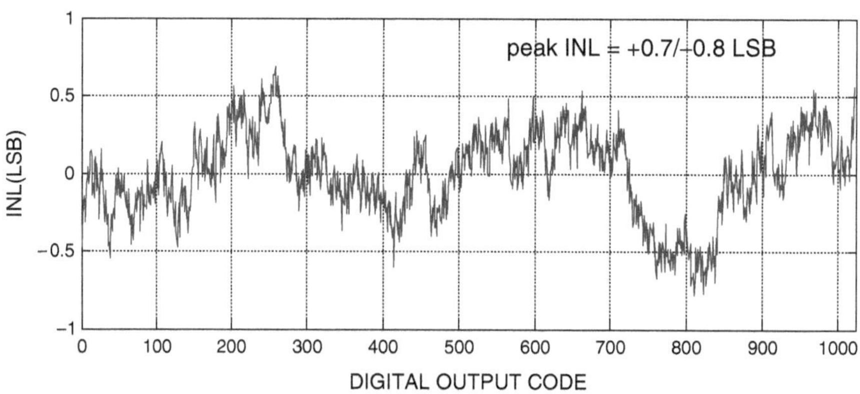

Fig. 9.48 INL after calibration (LSB @ 10-bit level)

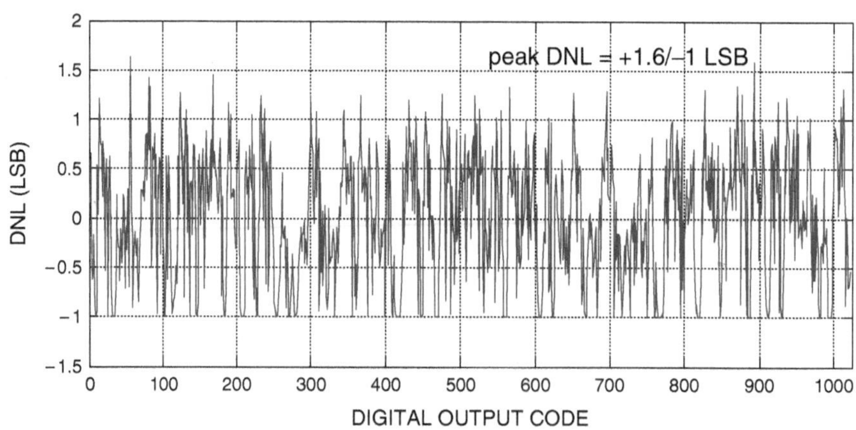

Fig. 9.49 DNL before calibration

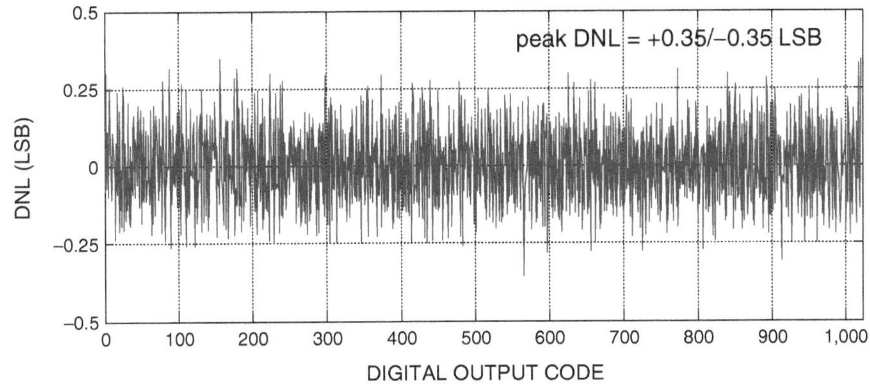

Fig. 9.50 DNL after calibration

Table 9.1 Summary of measured results

Technology	1.8 V, 0.18 μm CMOS
Input signal swing	1.0 V p-p
Area	1.4 mm^2
Sampling rate	50 MS/s
SNDR	58.2 dB (9.4b ENOB)
SFDR	66 dB
Power	9.9 mW
FOM	0.3 pJ/step

9.12.3 INL/DNL plots

Figures 9.47–9.50 show INL/DNL of the ADC before and after calibration, where it seen that calibration significantly improves the INL of the ADC from +15.7/−17.9 LSB to +0.7/−0.8 LSB and DNL from +1.6/−1 LSB to +0.35/−0.35 LSB.

9.13 Summary

In this chapter a technique to significantly reduce MDAC power was discussed. A differential charge sharing technique in combination with a simple source follower as a unity gain buffer and foreground calibration was shown to enable much lower power consumption over prior opamp based MDAC topologies. A summary of key measurement results is presented in Table 9.1.

Chapter 10
Summary

10.1 Summary

In this book a review of the fundamentals involved in designing a pipelined ADC was given. The key design tradeoffs in designing a pipelined ADC were detailed in Chapters 2–5, where the discussion provided an overview of the motivations of several state of the art enhancement techniques.

Chapter 6 dealt with a novel technique to enable rapid background calibration of both gain and DAC errors in a multi-bit pipeline stage using a dual-ADC based approach. Measured results from a 1.8 V, 0.18 μm CMOS prototype showed a more than $\times 100$ faster calibration than previous published reports, where a peak SNDR/SFDR of 60/70 dB was achieved in only 10^4 clock cycles, making the technique very useful in an industrial setting.

Chapter 7 described the implementation of a 10-bit 50 MS/s ADC which has its power scalable over a wide range of sampling rates – 35 mW at 50 MS/s and 15 μW at 1 kS/s, while maintaining an SNDR >54 dB over all sampling rates. The ADC was enabled by a Rapid Power-on Opamp topology which was able to power-on within a very short time interval.

Chapter 8 presented a technique where the front end sample-and-hold of a pipelined ADC for sub-sampling applications was eliminated, thereby enabling a large reduction in power consumption. Unlike prior published reports the technique did not require a carefully matched (and thus costly) layout to achieve functionality for very high input frequencies. Measured results from a prototype fabricated in a 1.8 V, 0.18 μm CMOS prototype showed better than 51 dB SNDR for input frequencies larger than 267 MHz.

In Chapter 9 a MDAC topology using a capacitive charge-pump based topology with digital calibration was used to implement a very low power pipelined ADC design *without* the use of power hungry opamps. Measured results of a prototype in 1.8 V, 0.18 μm CMOS showed the 50 MS/s ADC to achieve an SNDR/SFDR has high as 58.2/66 dB while only consuming 9.9 mW, yielding a figure of merit which is amongst the best 10-bit medium to high speed ADCs published.

I. Ahmed, *Pipelined ADC Design and Enhancement Techniques*,
Analog Circuits and Signal Processing,
DOI 10.1007/978-90-481-8652-5_10,

The following tables summarize the performance of the results presented in Chapters 6–9 (Tables 10.1–10.4).

Table 10.1 Summary of measurement results from Chapter 6

Technology		1.8 V, 0.18 μm CMOS
Sampling rate (f_s)		45 MS/s
Input signal swing		1.3 V p-p
Area		3.57 mm^2
Power		81 mW
Number of calibration cycles		10^4 cycles (0.22 ms)
	Before calibration	After calibration
SNDR	46.9 dB	60 dB
SFDR	48.9 dB	70 dB
INL	+6.4/−6.1 LSB	+1.1/−1 LSB
DNL	+1.1/−0.4 LSB	+0.45/−0.4 LSB

Table 10.2 Summary of measurement results from Chapter 7

Technology	1.8 V, 0.18 μm CMOS
Area	1.21 mm^2
Sampling rate (f_s)	<1 kS/s–50 MS/s
Input signal swing	1.6 V p-p
Power	15 μW (1 kS/s) – 35 mW (50 MS/s)
SNDR (f_{in} = 20.94 MHz)	55 dB (8.8 bit ENOB)
SFDR (f_{in} = 20.94 MHz)	67 dB
INL/DNL	+0.63/−0.91, +1.06/−1.2 LSB

Table 10.3 Summary of measurement results from Chapter 8

Technology	1.8 V, 0.18 μm CMOS
Area	1.21 mm^2
Sampling rate (f_s)	<164 kS/s–50 MS/s
Input signal swing	1.6V p-p
Power	<0.6–27 mW
SNDR	>51.5 dB for all f_s
SFDR	>60.5 dB for all f_s
Input frequency range	0–267 MHz
Power of Error! Reference source not found. @ 50 MS/s	35 mW
Power of this work at 50 MS/s	27 mW

Table 10.4 Summary of measurement results from Chapter 9

Technology	1.8 V, 0.18 μm CMOS
Input signal swing	1.0 V p-p
Area	1.4 mm^2
Sampling rate (f_s)	50 MS/s
SNDR	58.2 dB (peak)
SFDR	66 dB (peak)
ENOB	9.4 (peak)
Power	9.9 mW
FOM	0.3 pJ/step

References

1. Rusu, A., Rodriguez de Llera Gonzalez, D., Ismail, M.: Reconfigurable ADCs enable smart radios for 4G wireless connectivity. IEE Circ. Dev. Mag. **22**(3), 6–11, May–June (2006)
2. Ahmed, I., Johns, D.A.: An 11-bit 45MS/s pipelined ADC with rapid calibration of DAC errors in a multi-bit pipeline stage. Solid-State Circuits Conference, 2007. ESSCIRC 2007. Proceedings of the 33rd European, pp. 147–150, 11–13 Sept 2007
3. Ahmed, I., Johns, D.A.: An 11-Bit 45 MS/s pipelined ADC with rapid calibration of DAC errors in a multi-bit pipeline stage. IEEE Solid-St. Cir. **43**(7), 1626–1637 (2008)
4. Ahmed, I., Johns, D.: A 50 MS/s (35 mW) to 1 kS/s (15 μW) power scaleable 10b pipelined ADC with minimal bias current variation. Solid-State Circuits Conference, 2005. Digest of Technical Papers. ISSCC. 2005 IEEE International, pp. 280–598, 6–10 Feb 2005
5. Ahmed, I., Johns, D.A.: A 50-MS/s (35 mW) to 1-kS/s (15 μW) power scalable 10-bit pipelined ADC using rapid power-on opamps and minimal bias current variation. IEEE J. Solid-St. Circ. **40**(12), 2446–2455 (2005)
6. Ahmed, I., Johns, D.A.: A high bandwidth power scalable sub-sampling 10-bit pipelined ADC with embedded sample and hold. Solid-State Circuits Conference, 2007. ESSCIRC 2007. Proceedings of the 33rd European, pp. 159–162, 11–13 Sept 2007
7. Ahmed, I., Johns, D.A.: A high bandwidth power scalable sub-sampling 10-bit pipelined ADC with embedded sample and hold. IEEE J. Solid-St. Cir. **43**(7), 1638–1647 (2008)
8. Ahmed, I., Mulder, J., Johns, D.A.: A 50MS/s 9.9mW pipelined ADC with 58dB SNDR in 0.18 μm CMOS using capacitive charge-pumps. Solid-State Circuits Conference, 2009. ISSCC 2009. Digest of Technical Papers. IEEE International, pp. 164–165, 165a, 8–12 Feb 2009
9. Lathi, B.P.: Modern Digital and Analog Communication Systems. Oxford University Press, New York (1998)
10. Johns, D., Martin, K.: Analog Integrated Circuit Design. Wiley, New York (1997)
11. Analog Devices Inc, Kester, W.: The Data Conversion Handbook. Newnes: New York 2005.
12. http://www.maxim-ic.com/appnotes.cfm/an_pk/1080, 25 July 2008 9:39PM
13. Craninckx, J., Van der Plas, G.: A 65fJ/Conversion-step 0-to-50MS/s 0-to-0.7mW 9b charge-sharing SAR ADC in 90nm digital CMOS. Solid-State Circuits Conference, 2007. ISSCC 2007. Digest of Technical Papers. IEEE International, pp. 246–600, 11–15 Feb 2007
14. Agnes, A., Bonizzoni, E., Malcovati, P., Maloberti, F.: A 9.4-ENOB 1V 3.8μW 100kS/s SAR ADC with time-domain comparator. Solid-State Circuits Conference, 2008. ISSCC 2008. Digest of Technical Papers. IEEE International, pp. 246–610, 3–7 Feb 2008
15. Lewis, S.H., Fetterman, H.S., Gross, G.F., Jr., Ramachandran, R., Viswanathan, T.R.: A 10-bit 20-Msample/s analog-to-digital converter. IEEE J. Solid-St. Circ. **27**(3), 351–358 (1992)
16. Kwok, P.T.F. et al.: Power optimization for pipeline analog-to-digital converters. IEEE Trans. Circ. Syst-II: Analog Digi. Sig. Process. **36**, 549–553 (1999)

17. Cline, D.: Noise, speed, and power trade-offs in pipelined analog to digital converters. Ph.D. in Engineering thesis, University of California Berkeley (1995)
18. Chien, G.: High-speed, lower-power, low-voltage pipelined analog-to-digital converter. Masters of Science thesis, University of California Berkeley (1996)
19. Uyttenhove et al.: Speed-power-accuracy tradeoff in high-speed CMOS ADCs. IEEE Trans. Circ. Syst –II: Analog Digi. Sig. Process. **49**, 280–287 (2002)
20. Ahmed, I.: A power scaleable and low power pipeline ADC using power resettable opamps. Masters of Applied Science thesis, University of Toronto (2004)
21. Bult, K., Geelen, G.J.G.M.: A fast settling CMOS Op Amp for SC circuits with 90-dB DB gain. IEEE J. Solid-St. Circ. **25**, 1379–1384 (1990)
22. You, F., Embabi, S.H.K., Sanchez-Sinencio, E.: Multistage amplifier topologies with nested G_m-C compensation. IEEE J. Solid-St. Circ. **32**(12), 2000–2011 (1997)
23. Razavi, B.: Design of Analog CMOS Integrated Circuits. McGraw-Hill, New York (2000)
24. Sumanen, L. et al.: CMOS dynamic comparators for pipeline A/D converters. 2002 IEEE International Symposium on Circuits and Systems (ISCAS), vol. 5, pp. 157–160, May 2002
25. Yang, W., Kelly, D., Mehr, L., Sayuk, M.T., Singer, L.: A 3-V 340-mW 14-b 75-Msample/s CMOS ADC with 85-dB SFDR at Nyquist input. IEEE J. Solid-St. Circ. **36**(12), 1931–1936 (2001)
26. Dally, W., Poulton, J.: Digital Systems Engineering. Cambridge University Press, Cambridge (2001)
27. Hernes, B., Briskemyr, A., Andersen, T.N., Telste, F., Bonnerud, T.E., Moldsvor, O.: A 1.2V 220MS/s 10b pipeline ADC implemented in 0.13/spl mu/m digital CMOS. Solid-State Circuits Conference, 2004. Digest of Technical Papers. ISSCC. 2004 IEEE International, vol. 1, 15–19, pp. 256–526, Feb 2004
28. Gulati, K., Lee, H. -S.: A low-power reconfigurable analog-to-digital converter. IEEE J. Solid-St. Circ. **36**(12), 1900–1911 (2001)
29. Streetman, Ben G.: Solid State Electronic Devices. Prentice-Hall, Englewood Cliffs, NJ (1995)
30. Yodprasit, U. et al.: Efficient low-power designs using MOSFETS in the weak inversion region. 1998 IEEE Asia-Pacific Conference on Circuits and Systems, pp. 45–48 (1998)
31. Silveria, F., et al.: A g_m/I_D based methodology for the design on CMOS analog circuits and its application to the synthesis of a silicon-on-insulator micropower OTA. IEEE J. Solid-St. Circ. **31**, 1314–1319 (Sept 1996)
32. Enz, C.C., Vittoz, E.A.: CMOS low-power analog circuit design. Proceedings of IEEE International Symposium in Circuits and Systems *(ISCAS)*, Chap. 1.2, pp. 79–133, Tutorials (1996)
33. Binkley, D.M. et al.: A CAD methodology for optimizing transistor current and sizing in analog CMOS design. IEEE Trans. Comput. Aid. D. Int. Circ. Syst. **22**, 225–337, February 2003
34. Gerosa, A. et al.: Enhancing output voltage swing in low-voltage micro-power OTA using self-cascode. Electron. Lett. **39** (2003)
35. Crols, J., et al.: Switched-opamp: An approach to realize full CMOS switched-capacitor circuits at very low power supply voltages. IEEE J. Solid-St. Circ. **29**, 936–942 (Aug 1994)
36. Baschirotto et al.: A 1-V 1.8MHz CMOS switched-opamp SC filter with rail-to-rail output swing. IEEE J. Solid-St. Circ. **32**, 1979–1986 (1997)
37. Waltari, M. et al.: 1-V 9-Bit pipelined switched-opamp ADC. IEEE J. Solid-St. Circ. **6**, 129–134 (2001)
38. Farjad-Rad et al.: A low-power multiplying DLL for low-jitter multigigahertz clock generation in highly integrated digital chip. IEEE J. Solid-St. Circ. **37**, 1804–1812 (2002)
39. Leung, K.N., et al.: A capacitor-free CMOS low-dropout regulator with damping-factor-control frequency compensation. IEEE J. Solid-St. Circ. **38**, 1691–1702 (Oct 2003)
40. Minch, B.A.: A low-voltage MOS cascode bias circuit for all current levels. 2002 IEEE International Symposium on Circuits and Systems (ISCAS), vol. 3, pp. 619–622, May 2002

41. Arias, A. et al.: Design of a CMOS fully differential switched-opamp for SC circuits at very low power supply voltages. 2001 IEEE International Conference on Electronics, Circuits and Systems (ICESCS), vol. 3, pp. 1545–1548, Sept 2001
42. Abo, A.: Design for reliability of low-voltage, switched-capacitor circuits. Doctor of Philosophy in Electrical Engineering, University of California Berkeley (1999)
43. Linnenbrink, T.E. et al.: ADC testing with IEEE Std 1241–2000. 2001 Instrumentation and Measurement Technology Conference (IMTC 2001), vol. 3, pp. 1986–1991, May 2001
44. Galton, I.: Digital cancellation of D/A converter noise in pipelined A/D converters. IEEE TCAS-II **47**, 185–196 (Mar 2000)
45. Siragusa, E., Galton, I.: A digitally enhanced 1.8-V 15-bit 40-MSample/s CMOS pipelined ADC. IEEE J. Solid-St. Circ. **39**(12), 2126–2138 (2004)
46. Ray, S., Song, B.-S.: A 13-b linear, 40-MS/s pipelined ADC with self-configured capacitor matching. IEEE JSSC **42**, 463–474 (2007)
47. Chang, D.Y., Li, J., Moon, U.K.: Radix-based digital calibration techniques for multi-stage recycling pipelined ADCs. IEEE Trans. Circ. Syst. I **51**, 2133–2140 (2004)
48. Grace, C.R., Hurst, P.J., Lewis, S.H.: A 12b 80 MS/s pipelined ADC with bootstrapped digital calibration. In IEEE International Solid-State Circuits Conference (ISSCC) Digest Technical Papers, pp. 460–539, Feb 2004
49. McNeill, J., Coln, M., Larivee, B.: A split-ADC architecture for deterministic digital background calibration of a 16b 1 MS/s ADC. Solid-State Circuits Conference, 2005. Digest of Technical Papers. ISSCC. 2005 IEEE International, vol. 1, pp. 276–598, 6–10 Feb 2005
50. Moon, U.K., Song, B.S.: Background digital calibration techniques for pipelined ADC's. IEEE Trans. Circ. Syst. II **44**, 102–109 (1997)
51. Kwak, S.-U., Song, B.-S., Bacrania, K.: A 15-b, 5-Msample/s low-spurious CMOS ADC. IEEE J. Solid-St. Circ. **32**(12), 1866–1875 (1997)
52. Erdogan, O.E., Hurst, P.J., Lewis, S.H.: A 12-b digital-background-calibrated algorithmic ADC with -90-dB THD. IEEE J. Solid-St. Circ. **34**(12), 1812–1820 (1999)
53. Galton, I.: Digital cancellation of D/A converter noise in pipelined A/D converters. IEEE TCAS-II **47**, 185–196 (Mar 2000)
54. Li, J., Moon, U.K.: Background calibration techniques for multistage pipelined ADCs with digital redundancy. IEEE Trans. Circ. Syst. II **50**, 531–538 (2003)
55. Chiu, Y., Tsang, C.W., Nikolic, B., Gray, P.R.: Least mean square adaptive digital background calibration of pipelined analog-to-digital converters. IEEE TCAS-I **51**, 38–46 (2004)
56. Wang, X., Hurst, P.J., Lewis, S.H.: A 12-bit 20-Msample/s pipelined analog-to-digital converter with nested digital background calibration. IEEE J. Solid-St. Circ. **39**(11), 1799–1808 (2004)
57. Siragusa, E., Galton, I.: A digitally enhanced 1.8-V 15-bit 40-MSample/s CMOS pipelined ADC. IEEE J. Solid-St. Circ. **39**(12), 2126–2138 (2004)
58. Li, J., Ahn, G.-C., Chang, D.-Y., Moon, U.K.: A 0.9-V 12-mW 5-MSPS algorithmic ADC with 77-dB SFDR. IEEE J. Solid-St. Circ. **40**(4), 960–969 (2005)
59. Liu, H. -C., Lee, Z. -M., Wu, J. -T.: A 15-b 40-MS/s CMOS pipelined analog-to-digital converter with digital background calibration. IEEE J. Solid-St. Circ. **40**(5), 1047–1056 (2005)
60. Ray, S., Song, B.-S.: A 13-b linear, 40-MS/s pipelined ADC with self-configured capacitor matching. IEEE JSSC **42**, 463–474 (2007)
61. Mehr, I., Singer, L.: A 55-mW, 10-bit, 40-Msample/s Nyquist-rate CMOS ADC. IEEE J. Solid-St. Circ. **35**(3), 318–325 (2000)
62. Ali, A.M.A., Dillon, C., Sneed, R., Morgan, A.S., Bardsley, S., Kornblum, J., Wu, L.: A 14-bit 125 MS/s IF/RF sampling pipelined ADC with 100 dB SFDR and 50 fs jitter. IEEE J. Solid-St. Circ. **41**(8), 1846–1855 (2006)
63. Gulati, K., Peng, M.S., Pulincherry, A., Munoz, C.E., Lugin, M., Bugeja, A.R., Li, J., Chandrakasan, A.P.: A highly integrated CMOS analog baseband transceiver with 180

MSPS 13-bit pipelined CMOS ADC and dual 12-bit DACs. IEEE J. Solid-St. Circ. **41**(8), 1856–1866 (2006)
64. Anderson, M., Norling, K., Dreyfert, A., Yuan, J.: A reconfigurable pipelined ADC in 0.18 μm CMOS. Symposium on VLSI Circuits, 2005. Digest of Technical Papers, pp. 326–329, 16–18 June 2005
65. Audoglio, W., Zuffetti, E., Cesura, G., Castello, R.: A 6-10 bits reconfigurable 20MS/s digitally enhanced pipelined ADC for multi-standard wireless terminals. Proceedings of the 32nd European Solid-State Circuits Conference, 2006. ESSCIRC 2006, pp. 496–499, Sept 2006
66. Craninckx, J., Van der Plas, G.: A 65fJ/Conversion-step 0-to-50MS/s 0-to-0.7mW 9b charge-sharing SAR ADC in 90nm digital CMOS. IEEE International Solid-State Circuits Conference, 2007. ISSCC 2007. Digest of Technical Papers, pp. 246–600, 11–15 Feb 2007
67. Verma, N., Chandrakasan, A.P.: An ultra low energy 12-bit rate-resolution scalable SAR ADC for wireless sensor nodes. IEEE J. Solid-St. Circ. **42**(6), 1196–1205 (2007)
68. Hsu, C.-C., Huang, C.-C., Lin, Y.-H., Lee, C.-C., Soe, Z., Aytur, T., Yan, R.-H.: A 7b 1.1GS/s Reconfigurable time-interleaved ADC in 90nm CMOS. IEEE Symposium on VLSI Circuits, 2007, pp. 66–67, 14–16 June 2007
69. Andersen, T.N., Hernes, B., Briskemyr, A., Telsto, F., Bjornsen, J., Bonnerud, T.E., Moldsvor, O.: A cost-efficient high-speed 12-bit pipeline ADC in 0.18-μm digital CMOS. IEEE J. Solid-St. Circ. **40**(7), 1506–1513 (2005)
70. Hernes, B., Bjornsen, J., Andersen, T.N., Vinje, A., Korsvoll, H., Telsto, F., Briskemyr, A., Holdo, C., Moldsvor, O.: A 92.5mW 205MS/s 10b pipeline IF ADC implemented in 1.2V/ 3.3V 0.13μm CMOS. IEEE International Solid-State Circuits Conference, 2007. ISSCC 2007. Digest of Technical Papers, pp. 462–615, 11–15 Feb. 2007
71. Kim, H.C. et al.: A 30mW 8b 200MS/s pipelined CMOS ADC using a switched-opamp technique. ISSCC Digest Technical Papers, pp. 284–285, Feb 2005
72. Wang, R. et al.: A 3.3mW 12MS/s 10b Pipelined ADC in 90 nm digital CMOS. ISSCC Digest Technical Papers, pp. 278–279, Feb 2005
73. Cheung, V.S.L. et al.: A 1-V 10.7-MHz switched-opamp bandpass ΣΔ modulator using double = sampling finite-gain-compensation technique. IEEE J. Solid-St. Circ. **37**, 1215–1225, Oct 2002
74. Murmann, B., et al.: A 12-bit 75-MS/s Pipelined ADC using open-loop residue amplification. IEEE J. Solid-St. Circ. **38**, 2040–2050 (Dec 2003)
75. Fiorenza, J.K., Sepke, T., Holloway, P., Sodini, C.G., Lee, H.-S.: Comparator-based switched-capacitor circuits for scaled CMOS technologies. IEEE J. Solid-St. Circ. **41**(12), 2658–2668 (2006)
76. Li, J., Zeng, X., Xie, L., Chen, J., Zhang, J., Guo, Y.: A 1.8-V 22-mW 10-bit 30-MS/s subsampling pipelined CMOS ADC. IEEE Custom Integrated Circuits Conference 2006, pp. 513–516, 10–13 Sept 2006
77. Abo, A.M., Gray, P.R.: A 1.5 V, 10-bit, 14MS/s CMOS Pipeline analog-to-digital converter. IEEE, VLSI Circuits, pp. 166–169, June 1998
78. Baker, R.J.: CMOS Circuit Design, Layout, and Simulation. IEEE Press/Wiley Interscience, New Jersey (2005)
79. Maulik, P.C., Mercer, D.A.: A DLL-based programmable clock multiplier in 0.18-μm CMOS with -70 dBc reference spur. IEEE J. Solid-St. Circ. **42**(8), 1642–1648 (2007)
80. Lee, T.-C., Hsiao, K.-J.: The design and analysis of a DLL-based frequency synthesizer for UWB application. IEEE J. Solid-St. Circ. **41**(6), 1245–1252 (2006)
81. Park, J., Park, H.J., Jae-Whui, K., Sangnam, S., Chung, S.: A 1 mW 10-bit 500 KSPS SAR A/D converter. Proceedings of IEEE International Symposium on Circuits and Systems **5**, pp. 581–584, May 2000
82. Confalonleri, P., Zarnprogno, M., Girardi, F., Nicollini, G., Nagari, A.: A 2.7 mW 1 MSps 10 bit analog-to-digital converter with built-in reference buffer and ILSB accuracy

programmable input ranges. Proceedings of European Solid-State Circuits Conference (ESSCIRC), pp. 255–258, Sept 2004
83. Dorrer, L., Kuttner, F., Wiesbauer, A., Di Giandomenico, A., Hartig, T.: 10-bit, 3 mW continuous-time sigma-delta ADC for UMTS in a 0.12 μm CMOS process. Proceedings of European Solid-State Circuits Conference (ESSCIRC), pp. 245–248, Sept 2003
84. Fiorenza, J.K., Sepke, T., Holloway, P., Sodini, C.G., Lee, H.-S.: Comparator-based switched-capacitor circuits for scaled CMOS technologies. IEEE J. Solid-St. Circ. **41**(12), 2658–2668 (2006)
85. Wang, R., Martin, K., Johns, D., Burra, G.: A 3.3 mW 12 MS/s 10 bit pipelined ADC in 90 nm digital CMOS. In IEEE International Solid-State Circuits Conference (ISSCC) Digest Technical Papers, pp. 278–279, Feb 2005
86. Chang, D.Y., Moon, U.K.: A 1.4-V 10-bit 25 MS/s pipelined ADC using opamp-reset switching technique. IEEE J. Solid-St. Circ. **38**(8), 1401–1404 (2003)
87. Miyazaki, D., Kawahito, S., Furuta, M.: A 10-bit 30-MS/s low-power pipelined CMOS A/D converter using a pseudodifferential architecture. IEEE J. Solid-St. Circ. **38**(2), 369–373 (2003)
88. Yin, W., Jiang, J., Xu, J., Ye, F., Ren, J.: An Undersampling 10-bit 30.4-MSample/s Pipelined ADC. IEEE Asian Solid-State Circuits Conference, 2006 (ASSCC 2006), pp. 343–346, Nov 2006
89. Arias, J. et al.: Low-power pipeline ADC for wireless LANs. IEEE J. Solid-St. Circ. **39**(8), 1338–1340 (2004)
90. Xia, B., Valdes-Garcia, A., Sanchez-Sinencio, E.: A 10-bit 44-MS/s 20-mW configurable time-interleaved pipeline ADC for a dual-mode 802.11b/Bluetooth receiver. IEEE J. Solid-St. Circ. **41**(3), 530–539, Mar 2006
91. Ryu, S.T., Song, B.S., Bacrania, K.: A 10-bit 50-MS/s pipelined ADC with opamp current reuse. IEEE J. Solid-St. Circ. **42**(3), 475–485 (2007)
92. Lu, C.-C., Lee, T.-S.: A 10-bit 60-MS/s low-power CMOS pipelined analog-to-digital converter. IEEE Trans. Circ. Syst. II: Express Briefs [see also Circuits and Systems II: Analog and Digital Signal Processing, IEEE Transactions on] **54**(8), 658–662 (2007)
93. Quinn, P., Pribytko, M.: Capacitor matching insensitive 12-bit 3.3 MS/s algorithmic ADC in 0.25μm CMOS. Proceedings of the IEEE 2003. Custom Integrated Circuits Conference, pp. 425–428, 21–24 Sept 2003
94. Dickson, J.F.: On-chip high-voltage generation in MNOS integrated circuits using an improved voltage multiplier technique. IEEE J. Solid-St. Circ. **11**(3), 374–378 (June 1976)
95. Lee, H.S.: Fundamental limits and practical issues of analog-to-digital converters. ISSCC (2006) (short course)
96. Schreier, R., Silva, J., Steensgaard, J., Temes, G.C.: Design-oriented estimation of thermal noise in switched-capacitor circuits. IEEE Trans. Circ. Syst. I: Regular Papers [Circuits and Systems I: Fundamental Theory and Applications, IEEE Transactions on] **52**(11), 2358–2368 (2005)
97. Karanicolas, A.N., Lee, H.S., Bacrania, K.L.: A 15 b 1 Ms/s digitally self-calibrated pipeline ADC. IEEE International Solid-State Circuits Conference, 1993. Digest of Technical Papers, 40th ISSCC, pp. 60–61, 263, 24–26 Feb 1993
98. Lee, S.-H., Song, B.-S.: Simplified digital calibration for multi-stage analog-to-digital converters. IEEE International Symposium on Circuits and Systems, 1993 (ISCAS '93), vol. 2, pp. 1216–1219, 3–6 May 1993
99. Hadidi, K., Sasaki, M., Watanabe, T., Muramatsu, D., Matsumoto, T.: An open-loop full CMOS 103 MHz -61 dB THD S/H circuit, Proceedings of the IEEE 1998 Custom Integrated Circuits Conference, 1998, pp. 381–383, 11–14 May 1998

Index

T

U

V

W

FSC
www.fsc.org
MIX
Papier | Fördert gute Waldnutzung
FSC® C083411

Kolibri 360 GmbH
Ferdinand-Jühlke-Straße 7
99095 Erfurt, Deutschland
produktsicherheit@kolibri360.de